NATIONAL
GEOGRAPHIC

POCKET GUIDE TO THE

Birds

OF NORTH AMERICA

NATIONAL GEOGRAPHIC

POCKET GUIDE TO THE

Birds

OF NORTH AMERICA

LAURA ERICKSON
JONATHAN ALDERFER

NATIONAL GEOGRAPHIC

WASHINGTON, D.C.

CONTENTS

||||||||||||||||||||||||||||||||||||||||||||||||||||||||||||||||||||||||||||||||||||||||||||||

Invitation to Birding

"Earth's the right place for love," wrote Robert Frost. Earth is also the right place for birds. Of all the warm-blooded creatures on the planet, there exist roughly twice as many species of birds as mammals. Except for the ocean depths, birds can be found everywhere mammals live—and beyond.

People who take up bird-watching as adults are often amazed to discover just how many birds are out there. Most of us will spot no more than a fraction of the almost 10,000 bird species, but the species visiting your own backyard represent a good beginning. Indeed, the easiest place to watch birds is right at home, especially if you have a good assortment of native trees and shrubs, a bird feeder, and a birdbath. A surprising number of birds can even be attracted to an apartment balcony in the city. Start looking, and you will see: Birds are everywhere, a delight to behold, a pleasure to learn to identify.

Why Learn Bird Identification?

Recognizing birds may not be necessary to individual survival, but learning about the creatures with which we share this planet, especially our own little corner of it, enhances our daily lives, giving us grace notes of joyful recognition as we travel to new places or look out our own windows, whether the birds we see are comfortably familiar or exciting new discoveries.

Using your senses in new ways enhances your ability to detect birds, and as your awareness grows, your progress accelerates. People with vision or hearing impairments can be excellent birders, locating and identifying birds entirely by either sound or visual cues. Those of us who can use both senses can train our eyes and ears to more easily detect birds with practice. The more birds we see and identify, the simpler it is to find and identify more.

The male Scarlet Tanager, a vivid study in red and black, migrates through backyard trees in the East during April and May.

Where to Find Birds

+ Feeding stations and birdbaths can offer long, leisurely looks at songbirds.
+ Ponds and other park waterways attract waterfowl and gulls.
+ Many states publish a birding guide or birding trail map with directions to the best birding locales.
+ State and local bird clubs and ornithological societies offer free or inexpensive local trips guided by experienced birders to birding spots they know best.
+ Birding festivals and bird club conventions can take you farther afield, to find more exotic species with the help of expert guides.
+ Books such as National Geographic's *Guide to Birding Hot Spots of the United States* can give you travel ideas to broaden your birding knowledge.

Optics

Binoculars are an essential tool for bird-watchers, especially when you want to see details on birds beyond the window feeder. Optical quality improves with cost, so buy the best pair that you can comfortably afford. Ten-power binoculars have the best magnification but are the hardest to hold steady, provide less light than the same size model in a lower power, and have a smaller field of view, making it trickier to locate birds in the first place.

Binoculars by the Numbers	7X35	8X40	10X50
MAGNIFICATION	7x	8x	10x
DIAMETER OF OUTER LENS	35 mm	40 mm	50 mm
WEIGHT	light	medium	heavy
FIELD OF VIEW	widest	medium	narrowest
STEADINESS OF IMAGE	best	medium	most visible "shake"

Most birders find that seven- or eight-power binoculars offer more than enough magnification. Because compact models are lighter and often less expensive, they are a popular choice.

Binoculars can take some getting used to. If you've never used binoculars before, practice on nonmoving objects first. Sight an object with your eyes, then train yourself to lift the binoculars and point them in the same direction. When you spot a bird, keep your eyes on it as you lift the binoculars up to your eyes.

Birding festivals and other wildlife events offer opportunities to compare binocular models outdoors, so you can see which work best for your eyes, hand size, and other needs. Many organized field trips will also give you chances to see birds through a spotting scope before you buy your own.

Note: If you wear glasses to see distances, keep them on as you use your binoculars, but make sure that the eyecups are folded or pushed down; otherwise, keep them extended so they frame your eyes. The extra time you gain by not lifting your eyeglasses can be critical to getting a good look before a bird flies off.

Using This Field Guide

This book is a sampler of the rich array of North American birds. We have selected 160 species in total, including both the birds most commonly observed throughout North America and a few birds less easily seen but considered iconic, such as the Bald Eagle. We included species from almost every family, hoping to entice you to explore our continent's amazing bird life while helping you grow familiar with the bird families and taxonomy.

Here are a few pointers that will help you get the most out of this pocket field guide.

✛ Look up every bird you see. Many will fly away before you can identify them, but with practice, you'll become a quicker observer.

✛ Notice the length and shape of the bill, relative length of the tail, body posture, and markings on the wing and head. These features, along with behavior and geographical range, are the best clues for identifying a species.

✛ Use the Key Facts to focus on each bird's most important field marks—the physical characteristics distinctive to a particular species.

Bald Eagles seldom allow us to approach close enough to see their tongues or eagle eyes, but an adult's white head and tail feathers are diagnostic.

✦ Use the range maps to check on whether a species is likely to be found in your area.

✦ Keep your field guide handy. Pick it up now and then to scan a few pages, even if you're not looking up a specific bird. That way, you can become familiar with the species you are likely to see.

✦ Whenever you identify a new bird—and even when you encounter a familiar one—look it up in the pocket field guide. Look at the illustrations and read the text for pointers on all the important features that distinguish that bird species from others.

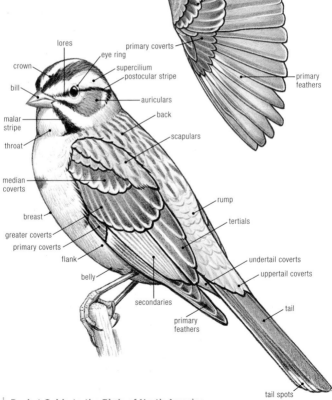

Identifying this Northern Mockingbird is just the first step. It's great fun to observe feeding behaviors and how they balance in tricky positions.

The more you use this field guide, the more proficient you'll be at finding new birds in it. By becoming familiar with it, you are learning the language of birders, and soon, even without this book in hand, you will have enough knowledge to place an unfamiliar bird in the right family and to notice details that can help you identify it later.

Taking It to the Next Level

Identifying birds is the first step. Watching behaviors, providing birdhouses, monitoring nest boxes, tracking migrations, noticing personality quirks in individuals of a single species—all these are the challenges and pleasures of continued bird-watching through the years. Many birders keep a life list—a record of the species they have spotted, as well as when and where. Photographing, videotaping, or drawing birds are ways of keeping a similar record. Joining in the birding community by uploading sightings to eBird.org is a great way to take your own observations to the next level, providing valuable data for researchers and helpful information for fellow birders.

Your experiences with birds, in the backyard and beyond, will connect you with nature in powerful, soul-satisfying ways. Come join us!

Canada Goose

Branta canadensis L 30–43" (76–109 cm)

Canada Geese were, until the 1970s, seen in many places only on migration and in winter.

KEY FACTS

Large dark goose; black neck; white chin strap; paler below.

+ voice: Call is deep but nasal *honk-a-lonk*.

+ habitat: Common and familiar. Feral birds frequent suburban areas, such as golf courses, parks, reservoirs; wild flocks migrate in V-formation.

+ food: Grazes on grass; eats a variety of pond life and waste grain in rural areas.

Geese use their feet to skid to a stop on water.

Geese are among a handful of birds capable of digesting grass, and so are drawn to expansive lawns. "Honkers" are extremely sociable. Once paired, they usually remain with their mate year-round for life. In winter, they may also maintain social bonds with young from previous seasons. They learn their migratory routes from their parents. When a pair of geese becomes urbanized and remains in a city for the winter, their young often follow suit. Adults undergo a flightless period while molting flight feathers in summer, before their goslings can fly. Goslings can swim and even dive to evade predators within hours of hatching.

honking calls

long, black neck and white chinstrap

adults

||||||||||||||||||||||||||||||||||||||||||||||||||||||||||||||||||||||||||||||||||||||||

Snow Goose
Chen caerulescens L 26–33" (66–84 cm)

White farm geese, domesticated in Eurasia, are
not the same species as wild American geese.

KEY FACTS

**Medium-size goose
with two color morphs:
White morph entirely
white with black wing
tips; blue morph has
dark gray body, white
head. Juveniles are
dingy gray.**

**+ voice: Gives single,
high-pitched calls.**

**+ habitat: Nests in Arc-
tic. Winters on marshes
and open fields, often in
large flocks.**

**+ food: Vegetarian;
eats various grasses,
weeds, and tubers.**

Flocks of Snow Geese can number in the millions.

"Waveys" got their nickname from their long, undulating
lines on migration. The two color morphs (white and
blue) were considered separate species until 1983. Birds of
both types usually select mates with the same plumage; the blue morph is dominant genetically, like brown
eyes in humans. These vegetarians are gregarious
except during the breeding season. Goslings remain
with their parents until their second or third
year. Populations are at historical highs,
and many biologists are concerned
about habitat destruction on the tundra
because Snow Geese grub for under-
ground tubers and roots, destroying
large swaths of delicate habitat.

black
wing tips

pinkish
bill

**blue-morph
adult**

**white-morph
adult**

Tundra Swan

Cygnus columbianus L 52" (132 cm)

Feral Mute Swans, brought here from Europe, displace native Tundra and Trumpeter Swans.

KEY FACTS

Larger and longer necked than geese. Entirely white; black bill with yellow spot near eye. Similar Mute Swan (not shown) has orange bill with black knob at base.

+ **voice:** Call is a single or double *honk* like an old car horn.

+ **habitat:** Nests in Arctic. Winters in large flocks in wetlands.

+ **food:** Aquatic plants and tubers, some mollusks, arthropods.

The serrated bill helps swans graze on plants.

Usually seen in flocks during migration and winter, America's "whistling swan" breeds in the Arctic and winters mostly near the Pacific and mid-Atlantic coasts of the United States. They once fed mostly on submerged aquatic vegetation, but habitat losses on their wintering grounds led them to depend increasingly on grain fields. Some wild Tundra Swans have lived over 23 years. They remain with their mate for life. Their chicks, called cygnets, can follow their parents within hours of hatching. Recently hatched cygnets are too buoyant to feed easily while swimming; they feed more often on land than the young of other swans.

pinkish bill

very long neck

juvenile

yellow spot in front of eye

adult

all white

Wood Duck

Aix sponsa L 18½" (47 cm)

One of the very few ducks that perches in trees and nests in tree cavities or nest boxes.

KEY FACTS

Glossy, colorful male is unmistakable; female is gray-brown with speckled breast, white belly. Juvenile similar to female.

+ voice: Female's flight call is loud squealing *oo-eek;* male gives a soft whistle.

+ habitat: Quiet rivers, wooded swamps, bottomlands; also along streams and rivers.

+ food: Aquatic plants and pond life; some acorns and grain.

Pairs stay together until all the eggs are laid.

This spectacular species declined dangerously in the late 1800s from overhunting for food and its beautiful feathers. Protection and nest box programs helped it recover. It takes two weeks or more for a female to lay about 13 eggs, but the entire clutch hatches out together within about 6 to 18 hours. The mother and chicks synchronize hatching by calling to one another while the ducklings are still inside the eggs. When it seems safe the morning after they hatch, the female calls to her brood from the ground, and one by one they jump safely down to her, as far as 50 feet below.

dark wings ♂

♀

longish tail

white around eye

♀

unique pattern

♂

Mallard

Anas platyrhynchos L 23" (58 cm)

Dabbling ducks are buoyant, with relatively
large wings. Most take off from water in a leap.

KEY FACTS

Male has green head
and yellow bill; female
is mottled brown with
darker bill.

+ **voice:** Female goes
quack-quack-quack;
male makes a raspy
kreep.

+ **habitat:** Abundant
and familiar denizens of
park ponds; wilder birds
live in a wide variety of
shallow-water wetlands.

+ **food:** Tips forward
to reach underwater
plant and animal life;
also forages on land.

Iridescent feathers change color as light changes.

The handsome Mallard, America's most abundant duck, is the parent species of most domesticated ducks. Farm ducks often mate
with wild Mallards, producing a variety of hybrids. If hybrids continue
to breed with wild Mallards, over generations, their off-
spring become more Mallard-like, though often larger
and with a wider white neck band. Each female
chooses her mate on her winter range; he follows her
back to where she originally hatched. Even devoted
males often mate with other females, so a
single brood of ducklings may have more
than one father. Mallards
have few taste buds on their
tongue, but many under the
tip of their bill.

pale
underwing ♂

blue
speculum ♀

dark saddle
on orange bill

green
head ♀

yellow
bill ♂

Northern Pintail

Anas acuta L 21" (53 cm) + 4" (10 cm) tail of male

Shallow wetlands from the prairies to the tundra are ideal breeding grounds for pintails.

KEY FACTS

Elegant and long necked. Male has brown head with white stripe, long tail; female is mottled brown with paler head (compare to female Mallard).

+ voice: Female *quacks;* male whistles.

+ habitat: Found in marshes and open areas with ponds; more common in West than East.

+ food: Tips forward to reach underwater plant and animal life; also forages on land.

Even brown feathers can be iridescent.

dark speculum
long pointed tail
white neck stripe
gray bill

The Northern Pintail is the greyhound of ducks—elegant, slender, and fast. From a distance, its delicate silhouette makes it seem smaller than it really is. Pintails nest almost as soon as ice goes out in northern areas. Each male remains with his mate until she has laid a full clutch and begins incubating. Nests may be a mile from the nearest pond or stream; as soon as ducklings hatch, they follow their mother to water. After recovering from botulism, one male banded in Utah in 1942 was found 82 days later in a flock of other exhausted pintails on Palmyra Island, an amazing 3,600 miles from where he was banded.

Green-winged Teal

Anas crecca L 14½" (37 cm)

Teal are highly sociable, short-necked dabbling ducks that fly very fast in large groups.

KEY FACTS

Small, compact, and fast flying; green wing patch; yellowish under tail. Male has chestnut head, green patch through eye, white shoulder bar; female mottled brown.

+ **voice:** Male call, a liquid *krick*; female, a shrill *quack*.

+ **habitat:** Aquatic habitats, including coastal marshes in winter.

+ **food:** Dabbles in shallows for seeds and small aquatic animals.

This courtship display shows off the male's colors.

Tiniest of our puddle ducks, Green-winged Teal have been called feathered minnows for their ability to twist, turn, and bank in high-speed flight and even dive into the water in tight formations in perfect unison. This is the only species of duck known to scratch itself while flying. The young can swim, dive, walk, and feed themselves within hours of hatching, but cannot regulate their body temperatures well when air temperature is less than about 50 degrees, so the mother may brood them both at night and during cool days for a week or more. The color teal was named for the shade of this bird's facial patch in some light.

flies fast

small size

♂

♀

green speculum

♀

chestnut head with green patch

small bill

♂

Ring-necked Duck

Aythya collaris L 17" (43 cm)

Diving ducks dive to feed and elude predators.
They run on the water's surface to take off.

KEY FACTS

Small diving duck with peaked head, bold white ring around bill. Male purple-black with pale gray flanks; female brown with gray face, white eye ring.

+ voice: Usually silent; soft growly calls.

+ habitat: Lakes and ponds; also coastal marshes, rivers in winter.

+ food: Dives for aquatic plants and insects; sometimes tips up like a dabbling duck.

The rusty neck ring can be hard to see.

The inconspicuous chestnut ring around the male's neck is visible only in certain light or at very close range, so hunters call this species the "ring bill." Like other diving ducks, ring-necks can be hard to track while they're actively feeding because they disappear from sight when they dive, often popping up some distance away. During migration, they often associate with other diving ducks, especially scaup, but the male's angular head, black back, and white vertical patch behind the black breast make this species easy to pick out. During the 1980s and early 1990s, when many other duck populations were declining, Ring-necked Ducks expanded their range.

black back

white ring on bill

peaked head

vertical white bar

Bufflehead

Bucephala albeola L 13½" (34 cm)

"Sea ducks" have dense plumage, extra fat reserves, and special glands to excrete salt.

KEY FACTS

Tiny diving duck with a large, puffy head. Very active when feeding. Male has large white patch on glossy black head; female has brown head with white cheek.

+ **voice:** Usually silent.

+ **habitat:** Fairly common; nests in woodlands near small ponds; in winter, found on sheltered bays, rivers.

+ **food:** Dives for insects, crustaceans, mollusks; some plants.

We identify many flying ducks by color patterns.

The Bufflehead takes its name, which means "buffalo head," from its large head and tiny body. It nests in abandoned flicker holes near freshwater lakes and rivers. Courting males make short flights over a female and land feet first, "skiing" on the water with white head feathers erect and bright pink feet fully exposed. Males frequently bob their head up and jerk it back in a dramatic courtship display. Unlike most ducks, pairs of Bufflehead sometimes return to mate year after year. These cavity nesters seldom walk on land except when females lead newly hatched ducklings to water. Hunters call them "butterballs" for their plumpness.

small bill

extensive white head pattern

♂

white cheek patch

♀

II

Common Merganser

Mergus merganser L 25" (64 cm)

Mergansers have narrow, serrated bills to grasp prey. Many nest in cavities and nest boxes.

KEY FACTS

Large diving duck with thick-based, red bill. Male has dark green head, bright white sides. Female has chestnut head, white chin, gray sides.

+ voice: Usually silent.

+ habitat: Prefers fresh water at all seasons; clear rivers and wooded lakes for breeding; winters on large rivers and lakes.

+ food: Pursues fish underwater; also mollusks, crustaceans.

Ducklings are the color of females.

These large, striking ducks often swim with their head under water up to the eyes as they watch for fish. They can slowly submerge without a ripple, or leap into a graceful dive. They prefer fishing in slow-moving, clear rivers and streams. Smaller, but shaped something like loons, males are easy to distinguish from loons by their color pattern, females by their "bad hair day" appearance. Males also have long head feathers but don't erect them. Chicks feed on aquatic insects for the first week or two, adding fish to their diet as they get bigger and more skilled. Few hunters target mergansers because of their strong fishy taste.

rufous head with white chin

red bill

green head

large size

white flanks

More Ducks

Over 40 species of ducks have occurred in North America, some of them rare visitors from other parts of the world. The 15 species shown here are common in North America and will give you an idea of the variety of colors and patterns possible. In winter, flocks of ducks are common. If you check closely, there's a good chance you'll see one of these species.

white of buffy white crown

♂

AMERICAN WIGEON

broken white eye ring

dark eye line

gray face

BLUE-WINGED TEAL

smaller bill than Cinnamon

♀

all but youngest males have red-orange eyes

CINNAMON TEAL ♀

♂

AMERICAN BLACK DUCK

streaked face ♀

NORTHERN SHOVELER ♀

orange line on bill

♂

RUDDY DUCK

rufous body ♂

bright blue bill

♀

single blurry line on face

GADWALL

mostly gray body ♂

even line of orange on sides of bill ♀

black undertail coverts

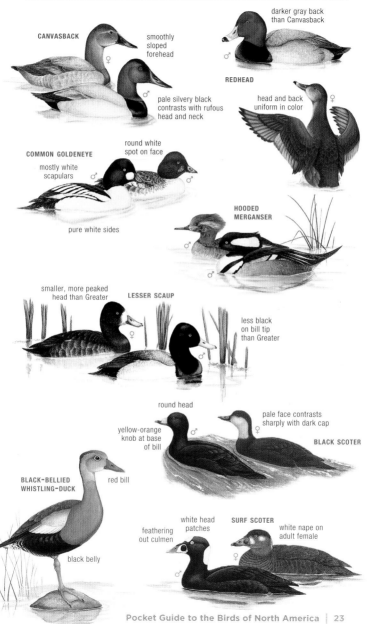

CANVASBACK — smoothly sloped forehead

darker gray back than Canvasback

♂

REDHEAD

pale silvery black contrasts with rufous head and neck

head and back uniform in color ♀

COMMON GOLDENEYE — round white spot on face

mostly white scapulars ♂

♂

pure white sides

HOODED MERGANSER

♂

♂

smaller, more peaked head than Greater

LESSER SCAUP

♀

less black on bill tip than Greater

♂

round head

yellow-orange knob at base of bill

♂

pale face contrasts sharply with dark cap ♀

BLACK SCOTER

BLACK-BELLIED WHISTLING-DUCK

red bill

black belly

white head patches

feathering out culmen

SURF SCOTER

white nape on adult female

♀

♂

California Quail

Callipepla californica L 10" (25 cm)

Quail aren't closely related to grouse and pheasants. They live in groups called coveys.

KEY FACTS

Ground-dwelling, chunky bird seen in groups; can burst into flight but usually runs for cover. Spiffy head pattern and topknot in male; female is plainer.

+ **voice:** Main call is loud *chi-CA-go*.

+ **habitat:** Brushy chaparral and scrubby lowlands in the West.

+ **food:** Vegetable matter and insects; mainly seeds and small fruits in winter.

The California Quail can run about 12 mph.

California's state bird is shy and retiring, more often seen in cartoons than in its natural range, but its loud call is often heard. A paired male and female alternate calls in a neatly arranged duet. Families usually associate in coveys—all the adults care for the young. This bird is well adapted to arid environments, and except during extreme heat can survive without drinking water, extracting fluids from food. It feeds mostly in the morning and evening, roosting in shade during daytime. When approached by a potential predator, it bursts into noisy flight.

smaller crest
brown face
teardrop-shaped crest ♀
lacks pattern **juvenile**
scaled belly ♂

|||||||||||||||||||||||||||||||||||||||||||||||||||||||||||||||||||||||||||||||||||||||||||||||

Northern Bobwhite

Colinus virginianus L 9¾" (25 cm)

The bobwhite, like the killdeer, phoebe, and chickadee, is named for the sound of its call.

KEY FACTS

Small chicken-like bird; only quail in East. Overall reddish brown with streaks and spots; male has boldly patterned head. Popular game bird.

+ voice: Loud whistled call *bob-WHITE!* is heard mostly in spring.

+ habitat: Farm fields and brushlands with good cover. Declining in most states.

+ food: Forages on ground for seeds and other plant material.

The male has a white throat and eyebrow.

Bobwhites are sociable quail the size of soccer balls. Males are aggressive toward other males during pair formation, but when both parents start incubating, males become peaceable again. Families often associate with other families soon after the fuzzy chicks have hatched. The species was once considered monogamous, but telemetry studies have revealed that both males and females may incubate and raise broods with more than one mate. Bobwhites were once very common, nesting in fencerows and other small patches, but are now declining in many areas. Captive-bred bobwhites are used to stock private hunting areas and to train retrievers. In many areas, most sightings are of escaped birds.

slight crest

white throat

buffy throat

♀

♂

Ring-necked Pheasant

Phasianus colchicus ♂L 33" (84 cm) ♀L 21" (53 cm)

The pheasant, one of the world's most popular game birds, is the state bird of South Dakota.

KEY FACTS

Long-tailed game bird introduced from Asia. Large, flashy male is iridescent bronze with fleshy red eye patches and white neck ring. Smaller female is buffy.

+ **voice:** Male's call a loud, harsh *kak-CACK!*

+ **habitat:** Resident in open country, farmland, brushy fields, woodland edges.

+ **food:** Forages on ground for seeds, grain, insects.

Bare facial skin is brightest when male displays.

Native to western Asia and eastern Europe, pheasants were traded and raised for food by ancient Greeks and Romans and introduced to many places over the world. Today, they sometimes escape from breeders and game farms even outside their range. As with other birds that mostly move about by walking or swimming, pheasant legs are powered by red muscle fibers—the dark meat—that are virtually tireless. The white muscle fibers powering their wings—the breast meat—produce rapid, strong bursts of flight, but tire quickly. Males may acquire a "harem" of several females.

white neck ring and red face

♂

long tail

iridescent bronze plumage

♀

buffy plumage

smaller than male

Wild Turkey

Meleagris gallopavo ♂L 46" (117 cm) ♀L 37" (94 cm)

Wild Turkeys are strong short-distance fliers, and usually roost in trees at night.

KEY FACTS

Very large game bird with iridescent, bronze plumage and bare-skinned head. Male larger than female with glossier plumage and breast tuft. Male displays in spring.

+ **voice:** Male's gobbling call may be heard a mile away.

+ **habitat:** Resident in open hardwood forest, woodland edges.

+ **food:** Forages on the ground for seeds, nuts, acorns, insects.

Displaying males are dominant over other males.

Audubon placed the turkey first in *Birds of America,* and Benjamin Franklin wanted it on our national emblem. Turkeys had been domesticated by the Aztecs, brought to Spain by conquistadors, and appeared on English tables by the 1540s. Back then, the wild birds fed on American chestnuts, walnuts, and other riches of the eastern forest. By the mid-1800s, turkeys had vanished from much of their range due to chestnut blight, deforestation, and overhunting. Now they've been successfully reintroduced over much of their former range and even beyond.

unfeathered reddish head

huge size

♂

♀

gray head

Ruffed Grouse

Bonasa umbellus L 17" (43 cm)

Ruffed Grouse may be red or gray, the color determined genetically like our hair color.

KEY FACTS

Chicken-like bird with mottled gray or reddish-brown plumage that matches the forest floor. Black ruff feathers and banded tail conspicuous only during display.

+ voice: Male's drumming sounds produced by rapidly beating wings.

+ habitat: Fairly common resident of deciduous and mixed woods.

+ food: Buds, twigs, leaves, nuts, berries.

Males erect the ruff (neck feathers) when drumming.

A deep, resonant *thump, thump, thump* starts up in an aspen forest. Like a bouncing ball, it speeds up until ending in a muffled roar. To make this drumming sound, the male Ruffed Grouse cups air in his "wing pits" and flaps hard. Females, unassisted by males after mating, produce up to 16 young in a single brood.

dark tail band

slight crest

black ruff

displaying ♂

♀

The chicks first fly when less than a week old. In winter, grouse feet grow comblike "snowshoes" that help them grip icy branches as they feed on aspen buds. Their intestines also change dramatically in winter to digest this woody tissue. The state bird of Pennsylvania is a popular game bird.

‖‖

Common Loon

Gavia immer L 32" (81 cm)

Sharp backward projections on upper palate
and tongue help loons hold slippery prey.

KEY FACTS

Large, low-riding waterbird. Breeding bird has checkerboard back, dark green head, striped collar; winter bird is dark gray with white throat and half collar.

+ voice: Iconic yodeling call heard mostly in breeding season.

+ habitat: Nests on northern lakes; winters in coastal waters and large ice-free lakes.

+ food: Mostly fish.

Loon with chicks adopts a defensive posture.

The state bird of Minnesota has oversize webbed feet that flatten on the forward stroke. These visual hunters choose lakes based on water clarity, and swallow prey headfirst. Their eyes are brightest red during breeding season. Loons nest at the shoreline or on floating islands and platforms. Their long, haunting "wail" is a contact call between family members. They make the laugh-like "tremolo" when stressed or in flight. Only territorial males make the nighttime "yodel" calls. They winter in salt water. A gland between their eyes excretes excess salt through their nostrils.

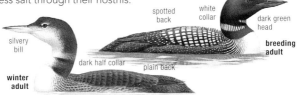

silvery bill

winter adult

dark half collar

spotted back

plain back

white collar

dark green head

breeding adult

Pied-billed Grebe

Podilymbus podiceps L 13½" (34 cm)

Grebes, nicknamed "hell divers" for their long dives, have lobed toes, not webbed feet.

KEY FACTS

Brown, duck-like, diving bird with lobed toes and thick, white bill. Black ring on bill and black throat; both lost in winter. Can hide by slowly sinking until only its head shows.

+ **voice:** Loud series of gulping notes heard on breeding grounds.

+ **habitat:** Nests around marshy ponds and sloughs; winters on fresh or salt water.

+ **food:** Dives for fish, crustaceans, insects.

Chicks keep safe between a parent's wings and back.

The Pied-billed Grebe is a tiny, somewhat secretive but widespread diving bird. It takes its name from the black-and-white, or "pied," bill of breeding adults. It may sink slowly into the water by expelling air from internal air sacs and from between its feathers, or may dive quickly. A nesting pair produces about four to six eggs. Zebra-striped chicks have red faces; they can swim soon after hatching but cannot maintain body temperature at first so often ride on their parents' backs. These nocturnal migrants are seldom seen in flight but have wandered as far as Hawaii and Europe.

black ring on bill

black chin

breeding adult

white throat

winter

downy young

Western Grebe

Aechmophorus occidentalis L 25" (64 cm)

Two similar grebes, Clark's and Western Grebes, were long believed to be the same species.

KEY FACTS

Large black-and-white grebe with long, thin neck and yellow-green bill. Very similar Clark's Grebe has more white around eye and orange bill.

+ voice: Call is a loud, two-note *crick-kreek*.

+ habitat: Nests in reed beds of fresh-water lakes; winters along Pacific coast and on large inland bodies of water.

+ food: Dives for fish and some crustaceans.

The rushing dance is beautifully synchronized.

This beautiful, slender-necked grebe is most famous for its "rushing" dance, when pairs or competing males rush toward each other and rise out of the water to run on the surface in synchrony. Mated pairs also engage in a "weed dance" during which each bird dives to grab vegetation, swims toward the other, and rises vertically out of the water in erect posture, bill lifted.

Western Grebes usually breed in colonies. They lay three to four eggs in a nest set in floating vegetation hidden among plants. They fly only during migration. After arriving on the breeding grounds, their flight muscles atrophy for the season, growing again in time for fall migration.

strikingly black-and-white head and neck; long, green-ish yellow bill

downy young

adult

Double-crested Cormorant

Phalacrocorax auritus L 32" (81 cm)

The double crest is held from spring to early summer, and varies from white to black.

KEY FACTS

Large, all-black water bird with yellow-orange facial skin; breeding adult has wispy crest. Young bird has pale breast. Often seen holding wings spread out to dry.

+ voice: Some grunting noises; usually silent.

+ habitat: Common and widespread along coasts, lakes, and rivers.

+ food: Pursues fish under water, propelled by large webbed feet.

Close up, cormorants are remarkably colorful.

Cormorants look like dark loons when swimming and like dark geese in flight. Wettable flight feathers decrease their buoyancy while diving. They are very sociable and nest in colonies with other cormorants and sometimes herons, on rocks and cliffs or in trees. All four toes are webbed. To incubate, they roll eggs atop their feet and rest the abdomen and breast on top of them. Hatchlings look like rubber toys, with naked dark brown, translucent skin. Within two weeks, they're covered with woolly black down. Clumsy on land, chicks and adults use their hooked bill tip to balance and walk.

yellow-orange face and throat skin

juvenile

pale breast

all-black plumage

winter adult

webbed feet

Anhinga

Anhinga anhinga L 35" (89 cm)

Nicknamed "snake birds," Anhingas often swim with only their neck and bill above water.

KEY FACTS

Long, snakelike neck and slender bill; long, broad tail. Male is black with silvery white patches on wings and upper back; female has buffy neck and breast.

+ voice: Mostly silent.

+ habitat: Prefers shallow, slow-moving waters and swamps. Often seen soaring high overhead.

+ food: Stalks fish under water, spearing them with its long, pointed bill.

Male Anhingas have dramatic breeding displays.

The only species in its family, the Anhinga's long tail counterbalances its long narrow neck and gives it a distinctive cross-like shape when soaring. Anhingas have dense bones and wettable feathers, allowing them to stay submerged while chasing and spearing fish. They lose body heat while under water, and must perch and spread their wings to dry their feathers and to warm their bodies in the sun. Heat loss during swimming restricts Anhingas to warmer climates than cormorants. They nest in loose colonies.

spears fish with long, pointed bill

silver spots on wings and upper back

buffy neck and breast ♀

breeding adult ♂

long tail

American White Pelican

Pelecanus erythrorhynchos L 62" (158 cm)

This huge pelican doesn't dive, but rather scoops up fish as it swims on the surface.

KEY FACTS

Huge waterbird with massive orange bill and throat pouch. White with black outer wing; juvenile with dusky marks on wings, neck, and head. Often soars.

+ **voice:** Mostly silent.

+ **habitat:** Breeds on inland lakes throughout the mountain West and northern Great Plains; winters along the coast.

+ **food:** Swimming birds corral fish, then dip their open bills to catch them.

Pelicans are known to swallow 3½-pound fish.

Pelicans are gangly on land, graceful in water and air. These gregarious birds hunt in groups, chasing fish to one another, herding schools of fish into shallow water, or encircling a large school of fish and closing in to concentrate them. Their lower mandible is flexible, widening enormously to serve as a scoop. This is the only pelican that grows a plate on the bright yellow upper bill during the breeding season. White Pelicans nest in large inland colonies. Both sexes build the nest on the ground and incubate the eggs under their huge webbed feet. Populations are growing, possibly due in part to fish farms.

nonbreeding adult

extensive black on wing

knob on bill

appears all-white when standing

orange feet and bill

breeding adult

Brown Pelican

Pelecanus occidentalis L 48" (122 cm)

The Brown Pelican is the only pelican species in the world that plunge-dives for food.

KEY FACTS

Very large, dark waterbird. Adult is gray-brown with blackish belly; top of head white to yellowish. Neck brown in summer, white in winter. Juvenile is dirty brown with pale belly.

+ **voice:** Mostly silent.

+ **habitat:** Saltwater. Flocks often seen gliding low over the water.

+ **food:** Plunge-dives for schooling fish and traps them in its ballooning pouch.

A pelican's pouch skin is extremely stretchy.

This sociable saltwater species often approaches people for handouts. In turn, gulls often grab fish from Brown Pelicans. Brown Pelicans fly in long lines barely above the water's surface, seeming to disappear behind large waves. They alight on water while extending their huge webbed feet as skids. A pelican's bill holds more than its belly can—after filling its pouch, it drains as much as 3 gallons of water before swallowing the fish. Brown Pelicans nest on the ground or in trees. The male selects the site and brings nesting materials; the female builds the nest.

dark head and neck

immature

pale belly

white neck

nonbreeding adult

red throat on West Coast

blackish belly

breeding adult

Great Blue Heron

Ardea herodias L 46" (117 cm)

Herons, unrelated to cranes, roost and nest in trees. Nest building can take up to two weeks.

KEY FACTS

Large, gray-blue heron; white crown with black stripe over eye; ornate head and neck plumes when breeding. Juvenile has black crown.

+ **voice:** Occasional deep croaks.

+ **habitat:** Common and widespread in fresh- and saltwater locations; also hunts in fields.

+ **food:** Includes fish, frogs, snakes, and small mammals.

These two nestlings won't leave the nest for weeks.

This huge bird has powerful neck muscles that drive its spear-like bill with amazing speed and precision. It swallows fish headfirst, usually waiting for larger fish and those with spines to die before swallowing. The bill and neck are relatively heavy, so as it takes off in flight, it kinks its neck to draw the weight nearer its wings. Like other herons, it does not carry prey, but feeds the young by regurgitating fish into the nest. Most herons nest in colonies (rookeries), but some pairs are territorial. Many select new mates each year, but some pairs reunite. In southern Florida, some are pure white or are pale with a white head.

flies with neck pulled in

black head stripe

adult

long legs

dark cap

juvenile

blue-gray plumage

breeding adult

Green Heron

Butorides virescens L 18" (46 cm)

This handsome, not-very-green heron's yellow legs turn orange during the breeding season.

KEY FACTS

Small, dark heron with short legs and dark bill. Adult has dark cap, chestnut neck, and greenish-blue back; juvenile is brownish above and streaked below. Usually solitary.

+ voice: Loud, sharp *kyowk*, often given when taking flight.

+ habitat: Wooded streams, ponds, and marshes.

+ food: Patiently stalks small fish and other aquatic life.

This heron hunts from many vantage points.

The Green Heron can stand absolutely still for several minutes along the water's edge until, with an explosive strike, it grabs or stabs prey. It sometimes baits fish, dropping bits of bread, insects, earthworms, twigs, or feathers onto the water's surface to lure them in. Like other herons, it sometimes fishes by night under moonlight. It usually nests in isolation but sometimes in loose colonies in swampy freshwater thickets, and often winters in coastal areas. Two nicknames, "shitepoke" and "chalkline," were given for its habit of conspicuously excreting a long stream as it takes off in flight. It's noisier than most herons, often squawking in flight.

adult

brownish above

short yellowish legs

streaked neck

juvenile

adult

small and dark

chestnut neck

Great Egret

Ardea alba L 39" (99 cm)

Egrets are herons with specialized nuptial feathers called aigrettes. Most are white.

KEY FACTS

Large white heron with heavy yellow bill, blackish legs and feet. In breeding plumage, long, wispy plumes trail from the back.

+ voice: Occasional deep croaking *kaaark*.

+ habitat: All types of wetland habitats, including damp fields.

+ food: Fish, invertebrates, reptiles, and small mammals. Walks slowly or waits for prey, then lunges with bill.

Displaying egrets are ethereally beautiful.

Great Egrets were hunted excessively for their beautiful breeding plumage, and became a symbol of the Audubon Society and conservation in general. They are widespread, found on every continent except Antarctica. They look huge but weigh barely 2½ pounds, less than half a Great Blue Heron's weight. They're aggressive toward other herons but surprisingly unaggressive toward potential nest predators—they've even been known to stand next to their nests without reacting as Fish Crows fly in and remove eggs. Some mooch for handouts at Florida theme parks. One lived over 22 years.

lime green facial skin

very long neck

long yellow bill

non-breeding

dark legs and feet

long plumes

high breeding adult

Snowy Egret

Egretta thula L 24" (61 cm)

The Snowy Egret's exquisite nuptial plumes are flaunted in spectacular breeding displays.

KEY FACTS

Small white heron with black bill and yellow feet ("golden slippers"). Breeding plumes on neck, head, and lower back.

+ voice: Low, raspy note, mostly near nest.

+ habitat: Found in salt- and freshwater habitats.

+ food: Small fish and crustaceans. More active than most herons; also uses feet to stir up prey in shallow water.

Heron bodies are deceptively small.

Snowy Egrets often run around in comically frantic pursuit of prey, wings held aloft. They are very sociable, often hunting in aggregations that include other wading birds, gulls, and terns, and nest in large colonies that usually include several species of waders. Both parents incubate the three to five eggs and feed and brood the young. After the breeding season, many Snowy Egrets wander well north of their breeding range. In 1886, this bird's plumes sold for $32 an ounce, double the price of gold at the time. The species declined until enactment of the Migratory Bird Act, which ended the feather trade.

adult

black bill and yellow feet

shaggy crest

high breeding adult

breeding adult

yellow-orange feet

Black-crowned Night-Heron

Nycticorax nycticorax L 25" (64 cm)

Night-herons hunt mostly at twilight and night.
By day, they are often seen roosting.

KEY FACTS

Stocky heron with short neck, big head, and yellow legs. Adult has black crown and back. Juvenile is brown with white spots on upperparts, streaks below.

+ voice: Loud, harsh *wok.*

+ habitat: Fairly common in salt- and freshwater wetlands, around ponds and nearby fields.

+ food: Fish, aquatic invertebrates, rodents, lizards, snakes.

Young night-herons are very different from adults.

These herons roost in trees or dense vegetation near water, even in large cities, and are one of the most widespread of all herons worldwide. Some winter in the north, feeding on mice when ice covers fishing areas. Like other herons, they nest in colonies, often sharing the same trees with other waders. Chicks leave the nest well before they can fly, climbing through branches and into other nests. They sometimes beg for and receive food from adults other than their parents. Black-crowns are often seen flying between feeding areas and roosts.

pale wings

adult

yellowish bill

juvenile

brown with white spots

black crown and back

white body

adult

White Ibis

Eudocimus albus L 25" (64 cm)

Ibises and spoonbills are large wading birds
with specialized bills. They are strong fliers.

KEY FACTS

White wading bird (not
a heron) with decurved
red bill, red face and
legs. Juvenile is dark
brown above; older
birds are mottled
brown and white.

+ voice: Occasional
low grunts.

+ habitat: Southern
swamps, coastal wet-
lands, beaches, lawns.
Often seen in flocks.

+ food: Probes
underwater for cray-
fish, crabs, and other
aquatic life.

This ibis's throat pouch is red during breeding.

White Ibises are abundant and conspicuous in the Deep
South, where they are mostly associated with marshes,
mangrove swamps, and other estuary and inland
wetlands. Foraging in shallow water, they use
their sensitive bill tip to locate prey by touch,
but when an injured ibis's bill tip is
replaced with a prosthetic, it can still
forage successfully. They also feed
on lawns and mooch for food at
Florida theme parks. They nest in
large colonies, which may lose up
to 44 percent of all eggs to predation. Independent young may join other
juveniles away from adults.

adults

black
wing tips

red face
and white
eye

white
body

dark
eye

juvenile

brown back
and white
belly

red legs

Roseate Spoonbill

Platalea ajaja L 32" (81 cm)

Spoonbills, unrelated to flamingos, are conspicuous waders that feed in tight flocks.

KEY FACTS

Flamboyant, pink wading bird with long, flat, spoon-shaped bill, unfeathered greenish head, white neck, and reddish legs. Juvenile much paler; takes 3 years to reach adult color.

+ **voice:** Mostly silent.

+ **habitat:** Fairly common in shallow water of swamps and marshes along Gulf Coast.

+ **food:** Mostly small fish, shrimp, and aquatic insects.

The chick's fuzzy head will be bald in a year.

Roseate Spoonbills feed by walking forward, sweeping their head side to side in wide semicircles, the sensitive bill slightly open until it detects prey items and snaps shut. Flocks may serve as "beaters," stirring up prey items for other waders that gravitate to them while feeding. They often roost in trees. When one flies over, a perched flock sometimes "sky-gazes," all extending their necks to point their bills skyward for a few seconds. They nest in colonies with other waders. Chicks insert their bill down a parent's throat to be fed. As they mature, they lose their head feathers and the bald skin gets more colorful.

unfeathered, greenish head

pink and red wings

pale pink wings

breeding adult

juvenile

spatulate bill

Wood Stork

Mycteria americana L 40" (102 cm)

America's only stork feeds by touch as well as sight, finding food easily in murky water.

KEY FACTS

Very large, white wading bird with heavy, downcurved bill and unfeathered gray head. Black flight feathers and tail; flies with neck extended, unlike herons. Juvenile has brown neck.

✛ **voice:** Mostly silent.

✛ **habitat:** Southern swamps, shallow pools, and coastal marshes.

✛ **food:** Mostly fish. Walks slowly through shallow water with open bill submerged.

Scalelike skin covers the back of the head and neck.

The "ironhead" or "flinthead" has been declining in South Florida since the 1960s, but small populations are expanding in the northern part of their range. Wood Storks nest in colonies with Anhingas and herons. To cool their chicks, parents regurgitate water over them. European storks nest on roofs near chimneys, perhaps why folklore credits them with delivering babies. Wood Storks weigh 4½ to 6 pounds, far too light to lug a newborn. Storks often spend hours on sunny days soaring on thermals.

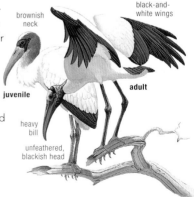

brownish neck

black-and-white wings

juvenile

adult

heavy bill

unfeathered, blackish head

Turkey Vulture

Cathartes aura L 27" (69 cm)

Vultures have distinctive bald heads, easy to clean after being plunged into carrion.

KEY FACTS

Blackish-brown with red head and pinkish legs; juvenile has grayish head, legs. In flight, note silver and black underwings and side-to-side rocking motion.

+ **voice:** Usually silent.

+ **habitat:** Common and widespread; favors wooded areas for breeding; open areas for foraging.

+ **food:** Animal carcasses, including roadkill.

Loose neck skin stretches to reach into carcasses.

Turkey Vultures locate carrion by scent. Power companies add ethyl mercaptan to natural gas; the smell attracts Turkey Vultures, which circle above ruptured pipes. Their gastric juices destroy disease organisms, but, given a choice, they prefer fresh food. They can devour skunks without piercing the scent glands. Their wing surface area is huge relative to their weight, so they can stay aloft on rising air currents with very little flapping. When trapped, they vomit toward potential predators, and urinate on their legs to cool themselves.

Osprey

Pandion haliaetus L 24" (61 cm)

Ospreys build stick nests at the very top of a
tree or power pole or on a nest platform.

KEY FACTS

Large, eagle-like
raptor. Brown above,
white below with
prominent dark eye
stripe. Juvenile marked
with pale fringes
above. Flies with
marked kink in wings.

+ voice: Series of loud,
whistled *kyew* notes.

+ habitat: Fresh- or
saltwater habitats with
clear water so that fish
are visible.

+ food: Large fish
caught by spectacular,
feet-first dives.

Chicks have dark, red-orange eyes; adult's are paler.

Osprey talons are designed for hanging on to slippery fish. One
front toe rotates backward to balance the two normal front
toes. Spiny scales under the toes make the hold even more secure.
When a fish is heavy, an Osprey may paddle to shore with
its wings. Bald Eagles sometimes steal fish
from them, one reason Ben Frank-
lin disapproved of
eagles. A courting
male performs a "fish
flight" sky dance.
Screaming and holding
a fish in his dangling
legs, he alternates hovering and steep ascents.

long angled
wings

bold dark
eye stripe

dark
"wrist"

adult

uniformly
dark above

prominent
pale tips

juvenile

adult

Mississippi Kite

Ictinia mississippiensis L 14½" (37 cm)

Kites are graceful, acrobatic hawks adapted for capturing small, erratically flying prey.

KEY FACTS

Graceful, mostly gray raptor with long, pointed wings and dark tail. Juvenile is heavily streaked below with barred tail and mottled underwings.

+ **voice:** Thin, high-pitched whistle.

+ **habitat:** Summer breeder in southern woodlands and shelterbelts; sometimes wanders north.

+ **food:** Mainly large insects; caught in flight.

Chicks are fed large insects, like this cicada.

This crow-size hawk with long, pointed wings can be mistaken for a falcon in fast flight. It's easier to recognize while soaring with wing and tail feathers spread or engaged in aerial acrobatics. Some nest in urban parks and golf courses where they defend nests aggressively, often joined by mockingbirds. Small songbirds sometimes mob them. Some build nests on wasp nests, which may protect them from climbing predators. The two chicks seldom squabble. Kites are most active at midday and late afternoon.

adult ♂

whitish patch

never hovers

juvenile

streaked underparts

banded tail

pale gray head

uniformly dark above

adult ♂

Northern Harrier

Circus cyaneus L 18" (46 cm)

Harriers, named for harrying prey, have owl-like facial discs to detect prey by hearing.

KEY FACTS

Long-winged, long-tailed raptor. Male is gray above, mostly white below; female is brown above, streaked below, with barred wings; juvenile resembles female but with rusty underparts.

+ **voice:** Various high-pitched calls.

+ **habitat:** Open wetlands, grasslands, pastures, and fields.

+ **food:** Flies low searching for rodents, reptiles, frogs.

Harriers scrutinize the ground while hovering.

Northern Harriers are buoyant, low-flying hawks that often hover while hunting. They sometimes follow large predators, seizing prey scared up by them, or along the edge of advancing wildfires, grabbing animals trying to escape. Adult males and females have different plumages. Females are mottled brown; males are nicknamed "gray ghosts" for their beautiful white and gray plumage accentuated by black wing tips. Males provide food and protection for their mate and young. In years of abundant prey, males may have a "harem" of 2 to 5 mates.

adult ♀

long tail

streaked belly

adult ♂

gray above

owl-like head

unstreaked belly

juvenile

Bald Eagle

Haliaeetus leucocephalus L 34" (86 cm)

Bald Eagles aren't really bald; the reference is to the adult's shining white head feathers.

KEY FACTS

Huge eagle. Adult has white head and tail, huge yellow bill. Juvenile is dark brown with mottled tail and underwings; birds attain adult plumage when about 5 years old.

+ **voice:** Series of high-pitched twitters.

+ **habitat:** Prefers seacoasts, large rivers, and lakes. Builds large stick nest in tall tree.

+ **food:** Mostly fish and waterfowl, also carrion.

Experienced adults catch more fish than immatures.

The Bald Eagle's imposing appearance led John Adams and Thomas Jefferson to use it on the national emblem, perhaps to symbolize America's military will and might. Eagle feet have three front toes and one rear toe. The imbalance of six toes on one side of prey and only two on the other may allow fish to thrash out of their grasp. In 1987, a Bald Eagle dropping its prey caused a midair collision between an Alaska Airlines jet and a fish; luckily, the only fatality was the fish. The oldest wild eagles known lived to their early 30s.

adult

dark head and bill

juvenile

white head

white tail

large yellow bill

adult

Cooper's Hawk

Accipiter cooperii ♂L 15" (38 cm) ♀L 18" (46 cm)

Accipiters have short, rounded wings and a long rudder-like tail to maneuver in woodlands.

KEY FACTS
Long-tailed, woodland hawk. Adult has dark gray upperparts, rufous bars below, and banded tail. Juvenile is brownish above, streaked below. Very similar Sharp-shinned Hawk is smaller.
+ voice: Nasal series of *kek* notes.
+ habitat: Woodlands, even in suburbs and towns.
+ food: Preys on small birds; often attacks birds at feeders.

This accipiter specializes in hunting birds.

Cooper's Hawks have become urbanized, nesting in large shade trees, but are secretive and seldom noticed outside migration except when hunting in backyards or defending a nest. Males build the nest and provide food for their mate from the time they begin courting through the time the chicks are 12 to 14 days old. Only females incubate eggs and brood and feed chicks. Females are significantly larger than males. They may chase prey on foot through dense undergrowth.

juvenile

large head and tawny nape

straight leading edge

long rounded tail

dark crown

juvenile ♀

adult ♂

smaller than female

Red-shouldered Hawk

Buteo lineatus L 17" (43 cm)

Buteos have long, broad wings and a short, broad tail and can float on rising air currents.

KEY FACTS

Medium-size hawk with distinctive black-and-white pattern on wings and tail, reddish bars on underparts, shoulders, and wing linings. Juvenile is browner with streaked breast.

+ voice: Loud, plaintive *kee-yeer*, often given in flight; imitated by Blue Jays.

+ habitat: Mixed woodland, particularly near water, swamps.

+ food: Small mammals, frogs, snakes.

These hawks eat many reptiles and amphibians.

This hawk has widely separated ranges. In the East, it is found in swamps, bottomland hardwood forests, and wet woodlands. In the West, it's found in riparian and oak woodlands, eucalyptus groves, and suburbs. It drops down from a perch to seize prey. Red-shouldered Hawks often mate for life. Many chicks are taken by Great Horned Owls. Turning the tables, one Red-shouldered Hawk chased a Great Horned Owl while the hawk's mate seized and ate the owl's chick. Migrating Red-shouldered Hawks join other hawks in large groups, called kettles, circling on the same thermals.

pale crescent

juvenile

adult

juvenile

barred tail

brownish above

streaked underparts

adult

red shoulders

black-and-white wings

Red-tailed Hawk

Buteo jamaicensis L 22" (56 cm)

In *Oklahoma*, the "hawk making lazy circles in the sky" is a buteo, most likely a Red-tail.

KEY FACTS

The "default" large hawk in North America, often seen soaring. Highly variable, but most adults have rufous tail and dark bar on front edge of underwings. Juvenile has banded (not rufous) tail and is heavily streaked below.

+ voice: Harsh, descending *kee-eerrrr*.

+ habitat: Varied, from mountains and forests to prairies and deserts.

+ food: Small mammals, reptiles, birds.

This is "Pale Male," Central Park's celebrated hawk.

Conspicuous Red-tailed Hawks live near humans even in large cities, and are also found in wilderness. They usually hunt from a perch until they spy prey, but can drop down to seize animals spotted while circling high above. Adults usually return to the same mate and nest year after year; nonmigratory pairs may remain on or near their territory year-round. Both adults incubate eggs and brood and feed the young. One famous Red-tail, "Pale Male" (see photo), constructed a nest on a building overlooking Central Park in New York City in 1992, and has raised young there for over two decades. Some wild Red-tails have lived over 30 years.

adult

pale breast

dark streaks on belly

dark patagial bar

broad wings

rufous tail

adult

American Kestrel

Falco sparverius L 10½" (27 cm)

Despite similarities to hawks, falcons are more closely related to parrots and songbirds.

KEY FACTS

Petite, colorful falcon. Male has bluish wings, barred chestnut back and tail, apricot underparts with black spots; female is reddish brown above with streaked underparts.

+ **voice:** Shrill, rapid *killy-killy-killy.*

+ **habitat:** Open country, farmland, roadsides. Nests in old woodpecker hole, birdhouse. Population is declining.

+ **food:** Mice, insects, and other small prey.

The male kestrel has very colorful plumage.

American Kestrels perch on power lines along highways; long wings make them seem to swell in size when taking off in flight. Kestrels often hover in place, studying the ground to find prey. They can see ultraviolet light, allowing them to detect urine-marked rodent trails. Unlike most raptors, males and females have different plumage, and females are larger and more aggressive. They nest in cavities—setting out nest boxes has been shown to help them spread into new areas. Nestlings shoot streams of droppings onto the walls of the nest cavity, giving the nest a strong odor of ammonia and probably killing bacteria.

frequently hovers

two facial stripes

adult ♂

bluish gray wings

adult ♀

rufous brown back and tail

adult ♂

|||||||||||||||||||||||||||||||||||||||||||||||||||||||||||||||||||||||||||||||||

Peregrine Falcon

Falco peregrinus L 18" (46 cm)

Peregrine Falcons can be found in almost all habitats on every continent except Antarctica.

KEY FACTS

Majestic, powerful falcon. Head blackish, appearing helmeted; blue-gray above; pale underparts with bars and spots. Juvenile brown above with heavily streaked underparts.

+ **voice:** Harsh *cack* notes around nest site.

+ **habitat:** Nests on tundra, coastal cliffs, also in cities; frequents open wetlands. Population is increasing.

+ **food:** Birds, hunted in flight at high speed.

Migrating peregrines may ride slowly on thermals.

The fastest known bird has been recorded flying over 200 mph in vertical dives, or stoops; up to 70 mph while directly chasing prey; and 30 mph while traveling along. Peregrines hunt nocturnal migrants at lighted skyscrapers and oil rigs. They were extirpated from the United States east of the Rockies by the 1960s, but reintroduction programs have been widely successful. Most Peregrines historically nested on cliffs, but a pair nested on the Sun Life Building in Montreal from 1936 until 1952, perhaps providing the original inspiration to build nest platforms on tall buildings.

dark moustachial stripe

brownish above

long wings

streaked underparts

juvenile

adult

More Raptors

Watching raptors is a very popular birding pursuit, especially during fall migration, when geography and prevailing winds funnel masses of birds past well-known hawk-watch locations such as Hawk Mountain (PA), Cape May (NJ), Hawk Ridge (MN), Corpus Christi (TX), and Marin Headlands (CA). Identifying flying raptors is a challenge—look for patterns on the underwings and tail. As you get more experienced, wing shape and flight style will help to identify distant birds.

BLACK VULTURE

whitish primary patch

CALIFORNIA CONDOR

white wing linings

adult

short tail

short tail

adult

WHITE-TAILED KITE

black patch

adult

long white tail

SHARP-SHINNED HAWK

short head projection

juvenile

adult

short tail with white band

streaked undertail

juvenile

COMMON BLACK-HAWK

NORTHERN GOSHAWK

broad wings

HARRIS'S HAWK
chestnut wing linings
adult
one black band

BROAD-WINGED HAWK
black trailing edge
banded tail
adult

GRAY HAWK
adult
black-and-white tail
gray underparts

long, narrow wings
juvenile
ZONE-TAILED HAWK
barred tail

SWAINSON'S HAWK
light-morph adult
dark chest band
thin, pointed wings

dark carpal patch
adult ♀
ROUGH-LEGGED HAWK
dark belly

FERRUGINOUS HAWK
adult
rufous leg feathering
whitish underwings

GOLDEN EAGLE
adult
dark body
dark underwings

Sora

Porzana carolina L 8¾" (22 cm)

Rails are secretive marsh birds that seem weak in flight but may migrate thousands of miles.

KEY FACTS

Most common rail in North America, but small and secretive. Thick yellow bill, black face and throat, streaked above. Juvenile lacks black feathering.

+ **voice:** Descending whinny and high-pitched *keek*.

+ **habitat:** Fresh and brackish marshes; saltwater marshes in winter.

+ **food:** Seeds and aquatic invertebrates.

Long toes support its weight on floating vegetation.

This common rail makes a distinctive whinny from within marsh vegetation. "Meadow chickens" provide food for harriers, coyotes, and other predators, and in fall, fattened on wild rice, they're hunted in 31 states and 2 provinces. To elude detection, these tasty, defenseless birds flatten their bodies laterally to slip between cattails without rustling them. A pair may produce up to 18 eggs in a single clutch. They build "dummy" nests to use as resting platforms or for brooding first-hatched chicks away from remaining eggs.

Soras have unwebbed feet but are fine swimmers, even right after hatching. The word *Sora* was derived from a Native American name for this bird.

buffy neck

thick yellowish bill

juvenile

chicken-like bird

black mask and throat

breeding adult ♂

American Coot

Fulica americana L 15½" (39 cm)

Coots and their relatives the gallinules are
often mistaken for ducks, but are not related.

KEY FACTS

Duck-like bird with
blackish plumage and
stubby white bill. Juve-
nile is paler below with
darker bill. Lobed toes
for swimming; requires
long, running takeoff.

+ voice: Grunting
and clucking calls.

+ habitat: Common.
Nests in freshwater
habitats; winters in both
fresh and salt water.

+ food: Aquatic
plants; often seen in
flocks, grazing (or loaf-
ing) on lawns.

Parents bring food to their chicks.

"Cute coots with their white snoots"
swim in flocks sometimes num-
bering in the thousands. They also feed on
land, usually in smaller groups. "Mud hens" are
awkward in flight, especially on takeoff—a flock
may noisily patter on water, beating their wings,
for a long distance before getting airborne.
Their long toes are lobed for swimming and also for supporting their
weight when standing in muck. The
lobes fold as the foot is lifted while walking
and in the forward stroke as they swim. Most birds in
this family have a tiny, triangular tail with fluffy white feathers
beneath. Some Bald Eagles specialize in hunting coots.

mostly
white bill

charcoal gray with
blacker head

adult

lobed
toes

swims like
a duck

Sandhill Crane

Grus canadensis L 41–46" (104–117 cm)

Cranes, a small family of large birds—15 species worldwide—are unrelated to herons.

KEY FACTS

Very tall, gray bird with long black legs and red crown. Flies with neck extended and has feather "bustle," unlike any heron.

+ voice: Very loud, bugling call *gar-oo-oo.*

+ habitat: Locally common. Breeds in marshes and on tundra; winter flocks forage in grain fields and wetlands.

+ food: Mostly grain and seeds, some insects.

Chicks stay close to parents for many months.

Huge numbers of Sandhill Cranes gather along Nebraska's Platte River in early spring to fatten up before completing spring migration. Cranes pair off when 2 to 7 years old, and remain together for life. Courtship dancing synchronizes their breeding readiness. Bare, bumpy skin on the crown gets more intensely red during territorial and courting activities. Chicks leave the nest soon after hatching. They remain with their parents for almost a year. Sandhill Cranes rub mud into their feathers while preening to stain them the color of local soil.

adult

red crown

neck extended in flight

tertial bustle, unlike Great Blue Heron

adult

stained adult

reddish stain from iron in mud

Black-bellied Plover

Pluvialis squatarola L 11½" (29 cm)

Plovers—short-billed shorebirds—run and pause while feeding. Some live far from water.

KEY FACTS
Large, chunky shorebird with short bill. Breeding adult has striking black face and belly; winter adult and juvenile are grayish above and streaked below.

+ voice: Rich, three-note whistle *pee-ooo-whee.*

+ habitat: Nests on Arctic tundra; winters on mudflats and beaches.

+ food: Marine invertebrates; mostly insects on tundra.

Black "wing pits" identify this plover year-round.

juvenile
black axillaries
white rump
winter
grayish pattern above
paler below
breeding ♂
black face and belly

Inconspicuous when not in breeding plumage, these handsome birds are among the most wide-ranging of all shorebirds. They winter along coasts of every continent except Antarctica and may appear just about anywhere during migration. Their large size and ability to change feeding methods help them adapt to extreme conditions; large eyes help them feed at night. Both parents incubate their four eggs in a lichen-lined ground nest in the high Arctic, and both raise the young. They call loudly at the first sign of danger, providing sentinel services for other shorebirds. Their rapid flight and wariness kept them common when market hunting decimated other shorebirds.

|||||||||||||||||||||||||||||||||||||||||||||||||||||||||||||||||||||||||||||||||||||||||||||||||||||||

Semipalmated Plover

Charadrius semipalmatus L 7¼" (18 cm)

This small wader can swim short distances, aided by partially webbed (semipalmated) feet.

KEY FACTS

Small, brown-backed shorebird with single black breast band and white underparts. Short, black-tipped, orange bill and orange legs.

+ **voice:** Unique, up-slurred whistle *chu-weet.*

+ **habitat:** Beaches, lakeshores, and tidal flats; seen across the continent on migration.

+ **food:** Small invertebrates found in wet areas.

Plovers spend most of the day walking and running.

This shorebird, often described as "adorable," is usually seen in flocks with other waders during migration. When not feeding near the water's edge on mudflats or sandy beaches, these plovers rest in small flocks safely above the high watermark. They have few defenses except subterfuge to protect their exposed ground nest, set in a shallow depression lined with debris, leaf litter, and other camouflaging items. Adults frequently sit low as if incubating eggs away from the nest, probably to confuse predators. They also lure away potential nest predators with a broken-wing display. They've recently expanded their Arctic breeding range southward.

juvenile

lacks black areas

darker legs

single breast band

yellow-orange legs **breeding ♂**

Killdeer

Charadrius vociferus L 10½" (27 cm)

The Killdeer was given both its name and its scientific name, *vociferus*, for its calls.

KEY FACTS

Large, dark plover with two black breast bands. Horizontal stance with long tail; orange rump (best seen in flight) and white eyebrow.

+ **voice:** Loud, piercing *kill-dee*.

+ **habitat:** Common in open fields (also lawns and parking lots) and on shores, but not tied to water.

+ **food:** Invertebrates, especially earthworms, grasshoppers, beetles.

Chicks huddle under their parent's warm feathers.

Our most widespread plover nests on the ground in short-grass meadows and burned-over tracts, and also on construction sites, gravel roads, and driveways, often far from water, sometimes even on flat rooftops. When a nesting Killdeer spots a potential predator, it performs an injury-feigning display. But when a grazing, nonpredatory animal such as a deer or cow approaches a nest, the Killdeer rushes toward it screaming loudly, which may turn the animal away. Chicks can see and follow their parents to food within hours of hatching. Killdeer follow plows to take churned-up worms and insects. They are fine swimmers.

orange rump

adults

two breast bands

long tail

American Avocet

Recurvirostra americana L 18" (46 cm)

Avocets belong to the family Recurvirostridae, the name referring to the up-tilted bill.

KEY FACTS

Black-and-white shore-bird with upturned bill (straighter in male) and long legs. Head and neck are rusty in breeding plumage, gray in winter.

+ voice: Loud, *kleek, kleek, kleek.*

+ habitat: Fairly common in shallow alkaline and saltwater wetlands.

+ food: Sweeps its long bill across the water (like a scythe) to catch small invertebrates.

Chicks stay close to a parent, but feed themselves.

Avocets sweep their bill through shallow water, detecting food by touch as well as sight. The bill is so sensitive that a bird in the hand recoils at the gentlest touch. Colonies breed in temporary wetlands of the arid West, where selenium contamination in irrigation water may cause embryo deformities. Incubating birds sit tight when aerial predators approach, but when off the nest aggressively strike at them. When a ground predator approaches, an avocet may fly at it while making changes to the pitch of its call. The Doppler effect makes the avocet seem to be approaching faster.

male has straighter bill than female

black-and-white body

cinnamon head and neck

breeding ♀

long gray legs

winter ♂

Spotted Sandpiper

Actitis macularius L 7½" (19 cm)

The "teeter-peep" has the most widespread breeding range of any American sandpiper.

KEY FACTS

Small sandpiper with a teeter-tottering walk; flies with stiff, stuttering wing beats. Breeding bird has spotted underparts and pink-orange bill. Juveniles and winter birds lack spots.

+ voice: Shrill *peet-weet*; in flight, series of *weet* notes.

+ habitat: Widespread in freshwater habitats, also seacoasts.

+ food: Probes and picks for invertebrates.

Breeding sandpipers sometimes sing from perches.

This common sandpiper is found along lakes and rivers, urban waterfronts, and agricultural ponds, bobbing its tail wherever it goes. It can dive underwater and swim or walk on the bottom to elude predators. When surprised by ground predators, it makes a display called a "rodent run," squealing as it crawls low to the ground, wings flapping and tail spread. Males usually provide more care for young than females do. The chicks stay with their father for at least four weeks, sometimes joining flocks with other families. Females often leave their chicks with their mate to start a new nest with another male.

lightly barred upperparts

juvenile

white below

breeding

heavily spotted below

teeters while walking

Greater Yellowlegs

Tringa melanoleuca L 14" (36 cm)

"Tattletales" are easy to find, making piercing alarm calls at the slightest disturbance.

KEY FACTS

Tall shorebird with long, bright yellow to orange legs. Breeding bird has streaked breast and barred flanks; juveniles and winter birds less heavily marked. Very similar to Lesser Yellowlegs but obviously larger (see photo).

+ **voice:** Strident series of *tew* notes.

+ **habitat:** Freshwater ponds and tidal marshes.

+ **food:** Small invertebrates, minnows.

Side-by-side Greater (left) and Lesser Yellowlegs.

Greater Yellowlegs, the size of small ducks, are widespread and easy to observe in wetlands during migration, and near coasts in winter. Small groups may work together to maneuver fish to where they can catch them easily. They spend as short a time as possible in the swampy muskeg habitats of Canada and Alaska where they breed. Little is known about their nesting habits because their breeding areas are so inaccessible and unpleasant, ridden with mosquitoes and black flies. Flukes (trematodes) infest Greater Yellowlegs so frequently and severely that parasites are a primary reason why they start migrating south in June.

winter

heavily
barred flanks

slightly
upturned
bill

whiter
below

breeding

bright yellow
legs

Willet

Tringa semipalmata L 15" (38 cm)

Willets breed in two very different locations: prairie wetlands and salt marshes.

KEY FACTS

Large, plump shore-bird. Plain and grayish overall, until it takes flight, revealing striking black-and-white wing pattern. Breeding bird is barred below; bars absent in winter.

+ voice: Loud, territorial call *pill-will-willet.*

+ habitat: Common along coasts in winter; breeds in Atlantic and Gulf salt marshes and western prairies.

+ food: Probes for invertebrates.

Willets are easy to recognize in flight.

Willets were named for their ringing *pill-will-willet* calls. Eastern and western populations vary in important ways. Eastern birds usually return to the same mate as long as both birds survive; western birds don't maintain pair bonds consistently. Eastern birds call more frequently and at a higher pitch, probably due to different ambient sounds. On their breeding grounds, Willets aggressively chase predators, but in winter they rise up in a flock to evade them. Scientists once placed them in their own genus, but DNA studies show they are fairly closely related to yellowlegs and some other sandpipers.

winter

striking wing pattern

large shorebird

heavy bill

plainer and paler

barred underparts

breeding

winter

Whimbrel

Numenius phaeopus L 17½" (45 cm)

Curlews, including the Whimbrel, are large
shorebirds with long, slender bills.

KEY FACTS

**Large, streaky brown
and buff shorebird
with long downcurved
bill. Bold dark stripes
on crown and through
eye; long grayish legs.**

+ voice: Fast series
of hollow whistles on
one pitch.

+ habitat: Nests on
Arctic tundra; along
coasts in winter and
during migration,
when found almost
worldwide.

+ food: Probes for
invertebrates.

The downcurved bill fits neatly into crab burrows.

This large wader's scientific name, *Numenius*, comes from Greek for "new moon," referring to the crescent shape of the bill, which matches the curve of fiddler crab burrows. On their wintering grounds, Whimbrels reach deep into these burrows for crabs. During breeding season and on migration they usually pick at berries, insects, and other food items with the bill tip. To swallow, they simply jerk the head back and catch the item in the back of the throat. Some Whimbrels make a nonstop flight from southern Canada or New England all the way to South America, a treacherous 2,500-mile journey. Pairs reunite on their breeding territory.

dark stripes
on head

adult

uniformly grayish
brown

long
decurved
bill

adult

dark legs

Ruddy Turnstone

Arenaria interpres L 9½" (24 cm)

Turnstones have a bill specially adapted for flipping stones to reveal food beneath.

KEY FACTS

Breeding male is striking, with black-and-white head, black-and-chestnut back, and short orange legs. Female is duller, and both sexes are less colorful in winter. Bold flight pattern.

+ voice: Low-pitched, guttural rattle.

+ habitat: Nests on Arctic tundra; along coasts in winter.

+ food: Aquatic invertebrates and insects.

The sharp, stout, upturned bill is a perfect tool.

Ruddy turnstones are exceptionally adaptable. Their short, sturdy legs and flattened toes, spiny beneath and armed with sharp claws, are adaptations for running on slippery wet rocks and holding their stance securely when prying into crevices and turning over stones and debris. During breeding season on the Arctic tundra, they feed almost entirely on flies. The rest of the year, they have an extremely varied diet that includes handouts from beachgoers. Breeding pairs are monogamous and territorial, but associate with other shorebirds elsewhere. They jab at birds approaching too close.

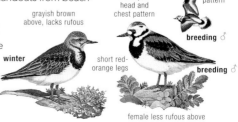

harlequin head and chest pattern

striking pattern

grayish brown above, lacks rufous

breeding ♂

winter

short red-orange legs

breeding ♂

female less rufous above

Sanderling

Calidris alba L 8" (20 cm)

These iconic sandpipers are found worldwide on temperate and tropical sandy beaches.

KEY FACTS

Small sandpiper. Winter plumage is very pale gray. Breeding bird is variably rufous. Bold wing stripe in flight.

+ voice: Series of *kip* notes. Flocks twitter.

+ habitat: Nests on Arctic tundra; along coasts in winter and during migration, when found almost worldwide.

+ food: Aquatic invertebrates and insects.

Sanderlings are usually seen in large flocks.

Sanderlings run ahead of advancing waves and chase retreating ones, giving them a frenzied, manic aspect. During the breeding season on the high Arctic, they are aggressively territorial. Either parent may incubate eggs and raise the young, but often only one does, the other finding a new mate and starting a new brood. Away from the breeding grounds, they associate in flocks that average 5 to 30 birds but may reach 2,500, especially along the Pacific coast. When chased by falcons, they fly in dense flocks moving erratically. When an individual is cut off from the flock, it may dive into the ocean.

variably rufous upperparts

black legs

breeding

very pale gray above

winter

feeds near breaking waves on sandy beaches

Least Sandpiper

Calidris minutilla L 6" (15 cm)

"Peeps" are small, similar-looking sandpipers that test birding skills.

KEY FACTS

The smallest of the small sandpipers, all of which are confusingly similar. The Least has yellow legs—unlike the others—and prefers to forage back from the water's edge rather than in the water.

+ **voice:** High *kreeep*.

+ **habitat:** Northern nester; freshwater and saltwater locations in winter and during migration.

+ **food:** Tiny invertebrates.

Diagnostic yellow legs may be covered with mud.

The tiny Least Sandpiper is often found with other peeps on mudflats, where muck can coat the diagnostic yellow legs. Quiet, slow-moving birders can make careful comparisons because they often allow close approach. Least Sandpipers return annually to the same territory on the subarctic tundra and northern boreal forest, and often reunite with the same mate as long as they both survive. The eastern population probably flies nonstop from the Northeast to South America, yet is surprisingly long-lived. One bird banded as a juvenile in Nova Scotia was recaptured there, alive, when 15 years old.

rufous fringes on back and wings

tiny sandpiper

heavily streaked breast

breeding

yellow legs

juvenile

lightly streaked breast with buffy wash

Wilson's Snipe

Gallinago delicata L 10¼" (26 cm)

Despite mythical "snipe hunts" with a paper bag and two sticks, the snipe is quite real.

KEY FACTS

Stock shorebird with very long bill, boldly striped head, barred flanks. Very well camouflaged; sits tight when approached, then bursts into flight, and flies off in a zigzag path.

+ voice: Harsh *ski-ape* call when flushed.

+ habitat: Wet meadows, bogs, and swamps for nesting; muddy fields during migration and winter.

+ food: Insects, worms, some seeds.

The snipe's bill has a sensitive tip for detecting prey.

The snipe is common and widespread but elusive and shy. In England, its name has been in use since at least the 1500s, probably a variant of *snout* in reference to the bird's long bill. Snipe probe deep into wet, mucky mud for worms and other prey that they grasp with the flexible tip while keeping the rest of the bill shut. Sharpshooting skills are essential to successfully hunt these erratic fliers, leading to the word *sniper*. Snipe make a variety of calls. The haunting winnowing sound, which serves as a territorial "song," is produced not by voice but by airflow over the outer feathers of the outstretched tail.

rapid, twisting flight

white stripes on back

head stripes

barred flanks

very long bill

||||||||||||||||||||||||||||||||||||||||||||||||||||||||||||||||||||||||||||||||||||

Laughing Gull

Leucophaeus atricilla L 16½" (42 cm)

Some "seagulls" live far from the sea, but the Laughing Gull is seldom far from salt water.

KEY FACTS

Midsize, dark gray gull, takes 3 years to become adult. Breeding adult has black hood and red bill; lacks hood in winter. First winter is browner and more subtly patterned, has black tail band.

+ voice: Crowing series of *hah* notes.

+ habitat: Common along Gulf and Atlantic coasts; rare inland.

+ food: Almost anything, including fish and beach handouts.

To mate, the male balances on the female's back.

Laughing Gulls are a familiar sight to anyone spending time in their range. They follow boats, mooch from picnickers, visit landfills, and grab fish from pelicans. Named for their flight calls, they are adapted for mobility on land, water, and air, often hovering to feed on flying insect swarms. Less aggressive than other gulls, they seldom take eggs or chicks from terns and shorebirds.

Early in the nesting season, large bathing groups may gather for 15 to 20 minutes of dunking and dipping. Chicks aren't buoyant and drown if high tides inundate the nest.

black hood

dark primaries

reddish bill

1st winter

breeding adults

brownish wings

black tail band

black legs

Ring-billed Gull

Larus delawarensis L 17½" (45 cm)

Many of these abundant, opportunistic gulls never spend any part of their lives on the sea.

KEY FACTS

Midsize, pale gray gull, takes 3 years to become adult. Adult has yellow bill with black ring, pale eyes, and yellow legs. First-winter bird has pink bill with dark tip, dark eyes, and brownish wings.

+ **voice:** Mewing *kee-ew;* sharp *kyow.*

+ **habitat:** Abundant and widespread along coasts and inland.

+ **food:** Almost anything, including fish and garbage.

Breeding adults have brilliant eyes and mouths.

This gull has adapted to human land-use patterns. Large flocks gather at dumps, freshly plowed or mowed fields, and fast-food restaurants. They nest on the ground near water in busy harbor and recreation areas as well as wilder spots. Their population expanded greatly between 1976 and 1990, especially in the Great Lakes region, and increased as much as 250 percent during the following two decades. In many areas, they're considered serious pests, but control measures have not curbed the population expansion.

It's hard to imagine today, but this gull was nearly wiped out by persecution and habitat destruction between about 1860 and 1920.

gray back

breeding adult

pale eye

1st winter

black band

pinkish legs

yellow legs breeding adult

||||||||||||||||||||||||||||||||||||||||||||||||||||||||||||||||||||||||||||||||||||

Herring Gull

Larus argentatus L 25" (64 cm)

Several large, white-headed gulls hybridize with
and can be confused with this common gull.

KEY FACTS

Large, pale gray gull,
takes 4 years to be-
come adult. Adult has
white head (streaked
in winter), yellow bill
with red spot, pale
eyes, pink legs. First-
winter bird is mottled
brown with dark bill
and pink legs.

+ voice: Loud *kyow.*

+ habitat: Wide-
spread along coasts
and inland.

+ food: Marine inver-
tebrates, fish, other
birds, carrion, garbage.

Parents take turns incubating for about a month.

Herring Gulls drink ocean water when no fresh water is available, their nasal glands excreting excess salt.
Away from the coast, they are most common around
large lakes and rivers. They catch wild natural
foods, scavenge on dead fish, and eat
garbage at landfills. To break open hard-
shelled prey, they often drop it on a rock or
other hard surface from the air. A Herring Gull in Europe was observed
using bread to bait goldfish.
Males regurgitate food to their
mate as part of courtship. Pairs
usually remain mated for life, remaining
with chicks for up to 6 months.

breeding
adult

1st winter

brownish
overall

dark tail

yellow bill
with red
spot

winter
adult

pink
legs

Caspian Tern

Hydroprogne caspia L 21" (53 cm)

The Caspian Tern is the largest and strongest of all terns, almost gull-like in flight.

KEY FACTS

Very large tern with a coral red bill and black undersides to wing tips. Breeding adult has a black cap, speckled in winter. Juvenile has dark edges to back and wing feathers.

+ **voice:** Loud, raspy *kowk* and *ca-arr*, often given in flight.

+ **habitat:** Breeds on large lakes (on islands) and in marshes; winters on southern coasts.

+ **food:** Mostly fish.

Adult carries fish crosswise to its nest.

Caspian Terns fly low over the water's surface searching for food, hovering and diving to take prey. They also snatch fish from smaller terns and gulls. Nesting Caspian Terns are fiercely protective, attacking hawks and inflicting bloody wounds on researchers' heads. Sometimes Herring Gulls and other opportunistic predators dart in and take chicks while the adults are chasing an eagle. Chicks need experience to get good at fishing. Meanwhile, their parents continue to feed them even on their wintering grounds. Some Caspian Terns "play," repeatedly dropping sticks or other objects and catching them before they hit the water.

breeding adult

black under-surface to primaries

solid black cap

streaked crown

thick red bill with dark tip

winter adult

large, size of a gull

Forster's Tern

Sterna forsteri L 14½" (37 cm)

Many terns are nicknamed "sea swallows" for their graceful flight and long, forked tail.

KEY FACTS

Midsize tern with a long tail; pale gray above, pure white below. Breeding adult has black cap and orange bill with black tip; winter adult has black face mask and dark bill.

+ voice: Hoarse *kyarr* notes, often in a series.

+ habitat: Fresh- and saltwater marshes, beaches, lakes, and rivers—the most likely tern in many places.

+ food: Mostly fish; some insects.

This tern may hover briefly before diving for fish.

This delicate-looking tern fishes in the deeper areas of fresh- and saltwater marshes, where it flies back and forth, bill pointed down, until it spots a fish and dives. One study found that about a quarter of Forster's Tern dives were successful. Small fish are swallowed almost immediately, but larger ones may be dropped a few times first. Unlike gulls, Forster's Terns virtually never swim. Despite their fragile appearance, they have harsh, raspy voices. They normally nest once per season, producing two to three eggs in a clutch, but renest if they lose their eggs or young to flooding or predation.

winter adult

overall very pale

black mask

breeding adult

long, forked tail

orange legs

Black Skimmer

Rynchops niger L 18" (46 cm)

Skimmers have distinctive, knife-thin bills, the lower mandible longer than the upper.

KEY FACTS

Large, black-and-white seabird with very long wings, a remarkable red-and-black bill, and short, red legs. Winter birds have a white collar. Juvenile's upperparts are fringed with white.

+ voice: Nasal *ip* or *yep.*

+ habitat: Marine areas with quiet water. Nests on sandy beaches.

+ food: Small fish caught by skimming.

The chick blends in well with its sandy habitat.

America's only skimmer is extremely sociable, nesting in colonies and associating in large flocks the rest of the year. When loafing on coastal beaches, skimmers, like other birds, invariably all face directly into the wind to keep air currents from ruffling their feathers. They usually forage singly, flying low, the knife-like tip of the lower bill skimming the water's surface. When it detects a prey item, the upper bill snaps shut. Detecting food by touch allows them to hunt at night under moonlight. When fishing by day, their slit-like pupils can be mostly closed to protect the eyes from harsh sunlight above and the glare of the water's surface below.

very long wings

black and white overall

breeding adult

"skimming"

||||||||||||||||||||||||||||||||||||||||||||||||||||||||||||||||||||||||||||||||||||

Atlantic Puffin

Fratercula arctica L 12½" (32 cm)

These penguin-like flying seabirds live on the open ocean except when breeding.

KEY FACTS

Compact seabird with a huge, multicolored bill, orange feet, and black-and-white plumage. Winter bird has dusky face and less colorful bill; juvenile's bill is much smaller.

+ **voice:** Growling notes at colonies.

+ **habitat:** Breeds in the North Atlantic, from Maine northward; winters at sea, a few as far south as Virginia.

+ **food:** Fish, pursued underwater.

Puffins use their webbed feet and wings underwater.

Every spring, these "clowns of the sea" return to the island where they were hatched to raise their own young. They dig a nest burrow using their bill and shovel-like webbed feet. After the single egg hatches, they feed the chick small fish, bringing about ten per foraging trip. To carry so many, they brace each one as it's caught against the spiny upper palate with their raspy tongue. After nesting, adults shed their bill's bright outer plates and the horny decorations above and below their eyes. Puffins were extirpated from many Atlantic islands; reintroduction efforts are bringing them back.

dark underwings

pale face

breeding adults

large colorful bill

red-orange legs

Pelagic Birds

These species spend their lives on the open ocean (pelagic) except when breeding. They include albatrosses, shearwaters, storm-petrels, tropicbirds, gannets, phalaropes, skuas, jaegers, alcids, and pelagic gulls and terns. All told, about 80 pelagic species have occurred in North American waters, so this is a small selection. Special offshore birding trips are the only way to see most of these species. Well-known locations that have regular boat trips include Cape Hatteras (NC), Monterey Bay (CA), and Westport (WA).

black bar

BLACK-FOOTED ALBATROSS
(Pacific)

white flash

BLACK-CAPPED PETREL
(Atlantic)

NORTHERN FULMAR

dark overall

very large

Atlantic light morph

PINK-FOOTED SHEARWATER
(Pacific)

pink feet

fluttery flight

long legs

WILSON'S STORM-PETREL
(Atlantic)

mottled underwing

GREAT SHEARWATER
(Atlantic)

MASKED BOOBY

adult

white rump band

white collar

pearly gray

black-and-white plumage

FORK-TAILED STORM-PETREL
(Pacific)

WHITE-TAILED TROPICBIRD (Atlantic)

adult

orange bill

black bar

RED PHALAROPE

winter

white stripe

black wing tips

BLACK-LEGGED KITTIWAKE

breeding adult

yellow bill

striking wing pattern

SABINE'S GULL

breeding adult

dark gray hood

ARCTIC TERN

breeding adult

short red bill

gray body

BRIDLED TERN (Atlantic)

breeding adult

white over eye

forked tail

POMARINE JAEGER

light morph breeding adult

white flash

breast band

long, twisted tail

COMMON MURRE

breeding adult

heavy body

black head

BLACK GUILLEMOT

breeding adult

oval white patch

Rock Pigeon

Columba livia L 14½" (37 cm)

"Dove" and "pigeon" are interchangeable. In common usage, "dove" refers to smaller species.

KEY FACTS

Pigeons come in many color varieties: pure white, checkered, tan, red-brown, and more. Typical birds are gray with a white rump and black wing bars.

+ voice: Throaty cooing.

+ habitat: Abundant in many cities, also around farmland.

+ food: Seeds, fruits, bread crumbs, and littered food.

Pigeons nest in building crevices and lay two eggs.

Pigeons were first domesticated for food, sport, and companionship at least 5,000 years ago. Their powerful homing instinct has been exploited to send messages in times of war and peace and for racing. Their intelligence and trainability make them valuable research subjects for studying navigation and orientation, physiology, and how animals learn. City pigeons in the United States are feral descendants of domesticated birds, including racing pigeons and even war heroes. Unlike most invasive species, pigeons don't compete with native birds, and provide food for urban Peregrine Falcons.

ancestral natural coloration

white rump

many other variations seen

color variations

Eurasian Collared-Dove

Streptopelia decaocto L 12½" (32 cm)

Released in the Bahamas in the mid-1970s, this dove soon reached the Florida peninsula.

KEY FACTS

Heftier and paler than Mourning Dove (next page), with a black half collar. Broad, squared-off tail is white with a black base. Perches in trees, overhead wires, rooftops.

+ voice: Three-syllable *coo-coo-cup.*

+ habitat: Starting from Florida, this nonnative species has colonized much of the country.

+ food: Seeds and grain.

This exotic species isn't found in older guides.

These handsome doves thrive in urban and suburban habitats and in rural places where grain is available. They often feed side by side with other ground-feeding birds, share roosts with native doves, and may nest near or even in the same trees

black collar

dark wing tips

overall pale gray-buff

as House Finches, House Sparrows, and, surprisingly, predatory Loggerhead Shrikes. Eurasian Collared-Doves are prolific, the female often laying new eggs as soon as still-dependent chicks fledge. Clutches usually have two eggs, and the first to be laid is significantly larger than the second. Pairs stay together for a full breeding season, and some remain together in winter and even through the following year.

Mourning Dove

Zenaida macroura L 12" (31 cm)

This dove was named for its mournful song, given most persistently by unmated males.

KEY FACTS

Slender dove with long, pointed tail. Rich tan above with black spots, paler below. When taking flight, its wings make a loud whirring noise and spread tail shows large white tips.

+ voice: Slow, owl-like *oowoo-woo-woo-woo.*

+ habitat: Abundant and widespread, except in dense forest.

+ food: Seeds and grain.

Doves drink water without lifting their bill.

About 1 million hunters shoot about 20 million Mourning Doves every year, more than all other game birds combined. Despite the harvest, Mourning Doves are surprisingly long-lived; some banded ones survived in the wild over 10 years and at least one over 31 years. They're monogamous during the breeding season. The female produces two eggs in a rather flimsy nest. Both parents incubate, and both feed the young "pigeon milk" produced in their crops. Due to their dry diet of seeds, Mourning Doves need a lot of water, especially when feeding young. Pioneers knew water was nearby when they spotted one.

long, pointed tail

black spotting on upperwing

scaly plumage

juvenile

slender body

White-winged Dove

Zenaida asiatica L 11½" (29 cm)

Some doves and pigeons have "orbital rings"—colorful patches of bare skin around the eyes.

KEY FACTS

Midsize, tropical dove with a prominent white wing patch. Orange-red eyes surrounded by blue skin.

+ voice: Drawn-out cooing *who-cooks-for-you.*

+ habitat: Found in woodlands, brush-lands, and desert towns in Texas and the Southwest.

+ food: Seeds, grain; also nectar, pollen, and cactus fruit.

This dove is well adapted to desert life.

The White-winged Dove, more easily heard than seen, has a call that is reminiscent of an owl. Enormous colonies, limited to dense woodlands in the Rio Grande Valley and Mexico, declined as natural habitat was developed for agriculture, grazing, and housing. It slowly adapted to more fragmented nesting habitat, shifting to backyard feeders and ornamental plantings, and now the population and range are expanding. An introduced population has also been established in Florida. All doves pick up grit on the ground to help digest seeds. When they ingest shot along with tiny pebbles on the ground, their bodies may accumulate toxic levels of lead.

red eye with blue orbital ring

crescent-shaped white wing patch

white edge to wing when perched

Monk Parakeet

Myiopsitta monachus L 11½" (29 cm)

North America's only *native* parrot, the
Carolina Parakeet, became extinct in 1918.

KEY FACTS

**Midsize parakeet with
bright green upper-
parts, pale gray face,
bluish wing tips, and a
long tail.**

+ voice: Loud,
grating squawks and
screams.

+ habitat: Wide-
spread, but decreas-
ing, in Florida; also
in scattered northern
cities.

+ food: Fruits, seeds;
also visits bird feeders.

These sociable parakeets sometimes visit feeders.

Monk Parakeets, commonly called "Quakers" by avi-
culturists, are the most widespread parrot
that escaped captivity to become established in
America. Native to temperate and subtropical
Argentina, they have adapted to winters as
far north as New York City, Chicago, and
Portland, Oregon. In South America, they
are considered agricultural pests. Fears of
ravaged crops and fruit trees in America
haven't materialized, and because so many
people are fond of these charismatic birds,
some eradication projects have been abandoned.
This is the only parrot that builds a stick nest. Colonies build
huge structures with individual apartments for nesting pairs.

bluish
underwings

green overall
with gray
forehead and
breast

long tail

|||||||||||||||||||||||||||||||||||||||||||||||||||||||||||||||||||||||||||||||||||

Yellow-billed Cuckoo

Coccyzus americanus L 12" (31 cm)

Although nicknamed the "rain crow," this bird's calls aren't correlated with precipitation.

KEY FACTS

Gray-brown above, white below, long blackish tail with large white spots. Yellow bill with black on tip and upper edge.

+ voice: Loud, hollow *kowlp-kowlp-kowlp*, staccato *kuk-kuk-kuk*, and series of *coo* notes.

+ habitat: Open woods, orchards, streamside groves. Rare in West.

+ food: Mostly caterpillars and large insects.

Cuckoos are usually secretive and hard to see.

These furtive birds are most easily seen near shrubs with webworms. American cuckoo calls include *coo-coo* notes, but none sound like cuckoo clocks, which were invented in the 18th century to mimic the Common Cuckoo of Germany's Black Forest. European cuckoos are brood parasites that lay their eggs in the nests of other species, leaving it to them to raise the young. American cuckoos raise their own young, though when food is abundant, Yellow-billed Cuckoos sometimes lay eggs in the nests of other cuckoos and species with blue-green eggs like theirs. In California, cuckoos sometimes breed cooperatively.

long tail

rufous primaries

adults

yellow bill base

large spots on black tail

Greater Roadrunner

Geococcyx californianus L 23" (58 cm)

Roadrunners make a series of slurred
coo-coo-coo notes, but never say *me-beep*.

KEY FACTS

Large, ground-dwelling
cuckoo streaked with
brown and white.
Bushy crest and heavy
bill. Very long, white-
tipped tail; short,
rounded wing with
black undersides.

+ **voice:** Dove-like
cooing.

+ **habitat:** Desert
scrub; less common in
chaparral.

+ **food:** Insects, liz-
ards, snakes, rodents,
and small birds.

It holds a scorpion by the tail before killing it.

New Mexico's state bird is perfectly adapted for desert life. It
gets enough water from its diet. When water is available, it
drinks with gusto but never bathes, preferring dust baths. It lowers
its body temperature at night, and raises it by sunbathing. It can fly
short distances but usually runs, maintaining speeds of 20
mph, using its tail as a rudder. It swallows horned lizards
headfirst with the backside up so the horns won't jab vital
organs; juveniles sometimes die when they swallow
one before learning this. When alarmed or territorial, roadrunners expose
orange skin behind
their eye—one cap-
tive bird fled when its
keeper wore orange.

crest

very long tail with
white tips

Barn Owl

Tyto alba L 16" (41 cm)

The only American bird in the family Tytonidae is nicknamed the "monkey-faced owl."

KEY FACTS

Very pale owl with dark eyes and a heart-shaped face. Tawny upperparts with white spots, white or buffy below. Strictly nocturnal; flies with slow wing beats.

+ voice: Harsh, hissing screech.

+ habitat: Open country with tree cavities, cliff crevices, or buildings to roost and nest in.

+ food: Small mammals, mostly rodents.

Owls listen for prey as they fly on silent wings.

heart-shaped face

dark eyes

females average darker below

long legs

Spooky hisses and raspy screams, ghostly white wings, and nests in steeples above churchyard graves probably contributed to folklore associating owls with death. Barn Owls also nest in barns, nest boxes, and natural cavities, and some nests can be observed via live Internet streaming. They're territorial around their nest but may share foraging territories with others. They usually mate for life. Some males roost near the nest; others use separate roosts. Broods average about five chicks; the more food available, the more survive to fledging. The Barn Owl's ability to locate prey by sound in absolute darkness is the most accurate of any animal tested.

Eastern Screech-Owl

Megascops asio L 8½" (22 cm)

In eastern woodlands, this small owl is often the most common avian predator.

KEY FACTS

Small, robin-size owl with a big head, ear tufts, and yellow eyes. Two color morphs: gray and rufous. Cryptic plumage blends in with tree bark of roost site (see photo).

+ voice: Two typical calls: a long trill and a quavering whistle.

+ habitat: Woodlands, suburbs, even large city parks.

+ food: Small animals, including birds, rodents, insects.

Cryptic coloration is this owl's primary defense.

Screech-owls spend their days in tree cavities and birdhouses year-round. If an intruder approaches one at the entrance to its cavity, it simply retreats. If the owl is on a branch, it stretches up and erects its feather tufts. If mobbed on a branch, it fluffs out, opens its eyes wide, and pulls back its tufts like an angry cat. Screech-owls usually select a mate of the same age, and pairs remain together for life. They often preen one another and their young, which reinforces their family bonds. They're very nurturing—in one case, a male Eastern Screech-Owl brooded flicker nestlings and tried to feed them bits of mouse.

ear tufts

pale greenish bill

gray morph

rufous morph

juvenile

more common in Southeast

Great Horned Owl

Bubo virginianus L 22" (56 cm)

The "feathered tiger" has the widest range and preys on more species than any American owl.

Not all owls that nest in barns are Barn Owls.

KEY FACTS

Very large owl with wide-set ear tufts. Overall color varies geographically from blackish brown (Pacific Northwest) to pale grayish (interior West) to reddish brown (East).

+ **voice:** Deep hooting *hoo hoo HOO-hoooo hoo.*

+ **habitat:** Widespread; prefers woodlands with open edges.

+ **food:** Mammals, birds, snakes.

The provincial bird of Alberta has a mellow hoot. Paired Great Horneds hoot back and forth, the pitch making it easy to tell which is which. Females are larger than males, but males have a deeper voice due to their larger syrinx (a bird's vocal apparatus). The heaviest ones weigh less than 4 pounds, yet have been documented killing prey as large as Great Blue Herons and Sandhill Cranes. There is little evidence that they can carry prey that big, but they often consume just the heads of larger prey, and eat where they made the kill, if left undisturbed. Pairs may remain together for life.

large ear tufts

yellow eyes

interior birds are paler and grayer

bulky body shape

Barred Owl

Strix varia L 21" (53 cm)

This brown-eyed owl, one of the most vocal owls in North America, often uses nest boxes.

KEY FACTS

Large, chunky owl with rounded head and dark eyes. Mostly brown with barred breast and streaked underparts.

+ **voice:** Deep hooting *who-cooks-for-you, who-cooks-for-YOU-ALL.*

+ **habitat:** Mature forest; common in river bottoms and southern swamps.

+ **food:** Small mammals, birds, reptiles, fish, and insects.

This owl's dark brown eyes are almost black.

Barred Owls are flourishing, expanding their range north and west, and even displacing endangered Spotted Owls. They mate for life and defend territories year-round, and may respond to imitations of their calls almost any time of year. Dueting Barred Owls often start with the familiar *who-cooks-for-you, who-cooks-for-YOU-ALL*, then move on to a series of ascending hoots ending with a loud *hoo-aw*, and finish off with a raucous jumble of calls sounding like maniacal laughter. Young ones sometimes make screechy begging calls all night long. Barred Owls live in swamps and wet forests, where they are known to pluck fish from ponds and streams.

dark eyes

barred breast

vertical streaks

Burrowing Owl

Athene cunicularia L 9½" (24 cm)

Unlike other owl species, male Burrowing Owls average slightly larger than females.

KEY FACTS

Brown with pale spots on back and brown bars on front. Glaring yellow eyes and long legs. Juvenile has buffy chest, no bars.

+ **voice:** Soft *coo-cooo* and chattering notes.

+ **habitat:** Open areas with short grass, prairie dog towns, airports, golf courses.

+ **food:** Insects, small mammals, birds, reptiles.

These owls eat insects and other small animals.

In the West, this popular little ground-nesting owl usually nests and roosts in burrows dug by prairie dogs, skunks, tortoises, and so on. Some use culverts. The isolated Florida population is more likely to excavate their own burrows. Burrowing Owls tolerate higher levels of carbon dioxide than do other owls, an adaptation for living in long burrows with little air exchange. Females incubate the eggs and brood and feed the chicks; males hunt. They place animal dung around the entrance and within the nest. This may attract dung beetles, which they eat, but also may help control climate and carbon dioxide levels within the burrow.

adult
spotted above
round head
no barring below
juvenile
long legs

Common Nighthawk

Chordeiles minor L 9½" (24 cm)

Nighthawks have such erratic, bat-like flight that they are nicknamed "bullbats."

KEY FACTS

Floppy flight style on long, swept-back wings that have a white bar across the wing tip. Cryptic plumage makes it hard to see when perched.

+ voice: Nasal *peent* call is unique.

+ habitat: Woodlands, suburbs, and towns; declining in the East. Migrates to South America.

+ food: Insects caught in flight.

Nighthawks have tiny feet and a huge mouth.

Unrelated to raptors, nighthawks feed only on the wing, darting about in the evening sky, capacious mouths wide open to swallow moths and other flying insects whole. They fly low over water to drink. The tiny bill is loosely attached and the tongue a vestigial flap in the back of the mouth; grounded nighthawks cannot pick up food. They used rock-ballasted flat roofs for nesting, and their *peent* calls and booming breeding displays were once common evening sounds in both cities and wild open habitat. Changes in roof construction and increases in urban gulls and crows contributed to their decline over much of their range.

long wings with white bar

♂

white throat

♂

juvenile

reduced white in wings

Eastern Whip-poor-will

Caprimulgus vociferus L 9¾" (25 cm)

In 2011, the Whip-poor-will was split into two
species: Eastern and Mexican.

KEY FACTS

Nocturnal and rarely
observed, but often
heard. Cryptic plum-
age makes it hard to
see when perched.
Large head and tiny
bill, but very large
mouth. Male's tail has
large white tips, buffy
in female.

+ voice: Loud, clear
whip-poor-will.

+ habitat: Open
woodland.

+ food: Flying insects;
active at dusk, dawn,
and on moonlit nights.

Big eyes help them track insects in the night sky.

Like nighthawks, whip-poor-wills belong to the nightjar family;
unlike nighthawks, their mouths are bordered by long, stiff feathers called rictal bristles. Males produce their famous call from a variety
of perches in their territory. The call attracts females but doesn't seem
to elicit aggression from nearby territorial birds. They don't build a
nest but, rather, lay their two eggs directly on leaf litter on
the forest floor. Both parents incubate the
eggs, brood the chicks, and feed
them regurgitated insects. One
study found that hatching
occurs about 10 days before
a full moon, allowing maximum
nighttime brightness for hunting
while the chicks are most rapidly growing.

often closes
eyes when
perched

male has
whiter tail

cryptic
plumage

Chimney Swift

Chaetura pelagica L 5¼" (13 cm)

Strong claws and stiff tail spikes allow swifts to roost on vertical structures.

KEY FACTS

Only swift in the East. Sooty brown, cigar-shaped bird, almost always seen in flight. Compared to a swallow, the swift has longer, swept-back wings and flies faster with stiffer wing beats.

+ voice: High-pitched, *chip* notes given in flight.

+ habitat: Most common in areas with chimneys for nesting.

+ food: Flying insects.

Swift saliva glues its twig nest to the chimney.

On summer evenings, Chimney Swifts flutter in the sky, making chittering calls as they swarm and funnel into communal roosts. Before chimneys were available, they used hollow trees and large woodpecker holes. Now, chimneys are often capped and lined with metal; some people provide roosting towers as a substitute. Only one pair of swifts nests in a single structure, though others may roost in it with them. They break off sticks with their feet and carry them in their bill, using gluey saliva to hold the nest together and affix it to the substrate (see photo). The nest of an Asian species—the nest used in bird's nest soup—is built entirely from viscous saliva.

swept-back wings

sooty brown plumage

short tail

cigar-shaped body

Ruby-throated Hummingbird

Archilochus colubris L 3¾" (10 cm)

Hummingbirds, requiring protein as well as carbs, are skilled at capturing tiny insects.

KEY FACTS

Only hummingbird in the East, Ruby-throat has glittering green upperparts. Male has red throat that looks black in some light; female's throat is white.

+ **voice:** Twangy *chips, tchew* or *chih.*

+ **habitat:** Forest edges, flowering gardens, sugar-water feeders.

+ **food:** Flower nectar, small insects, spiders.

This hummer's wings flap 53 beats per second.

Weighing less than a nickel, many Ruby-throated Hummingbirds fly nonstop over the Gulf of Mexico to the Yucatán Peninsula, a journey of at least 500 miles; in fall, they make this flight during hurricane season. They dive-bomb birds as large as Bald Eagles.

Hummers arrive in spring before flowers open; they can feed on sap and insects at sapsucker drill holes (page 101). Nests are tiny and well insulated to hold the mother's belly tight against her pea-size eggs without letting her warmth escape. To stretch as chicks grow, the nests are made of bits of lichen woven with spider silk, which also holds the nest to the branch.

ruby red throat

adult ♂

forked tail

♀

bright green above

whitish throat

prominent tail spots in flight

Anna's Hummingbird

Calypte anna L 4" (10 cm)

The only nonmigratory U.S. hummingbird; wanders widely after nesting season.

KEY FACTS

Metallic green above and dingy grayish white and green below. Male has rose red throat and crown; female's throat is white with spotting.

+ **voice:** Male's song is a rhythmic series of scratchy notes.

+ **habitat:** Common around backyard gardens and flowering ornamental trees. Year-round resident.

+ **food:** Flower nectar, small insects, spiders.

This nest stretches to fit growing chicks.

The entire head of a male Anna's Hummingbird glitters like a sparkling amethyst, and he sings a complex, learned song as well as performing a flight display to attract females. At the bottom of his display dive, he makes a complex sound produced in part by the wings and in part by the spread outer tail feathers. Flight muscles comprise 28 percent of this hummer's body mass. Anna's Hummingbirds begin nesting at the onset of winter rains except in the more northern reaches of their range. The species has been expanding its range eastward, taking advantage of feeding stations and garden flowers.

rose red head

adult ♂

rather short bill

adult ♀

usually with some red on throat

dingy gray below with green flanks

Broad-tailed Hummingbird

Selasphorus platycercus L 4" (10 cm)

This tiny hummingbird's wings produce a loud trilling sound rather than a hum.

KEY FACTS

Tail is impressively large. Male has a rose red throat; female has a white throat with tiny green spots and buffy underparts.

+ **voice:** Metallic *chip* notes; adult male's wings make a cricket-like trill in flight.

+ **habitat:** Summer resident of foothills and mountain meadows. Winters in Mexico.

+ **food:** Flower nectar, small insects, spiders.

Trilling wings draw our attention to this bird.

Broad-tails nest at high elevations in the central and southern Rocky Mountains. They can allow their body temperature to drop from a normal active temperature of 100° to 109°F down to as low as 54°F when it's cold. In forward flight, their wings beat about 38 times per second; about 50 times per second while hovering to feed. As with other hummers, they don't form pair bonds—one male was reported mating with at least six different females. This species holds the North American hummingbird longevity record: A female caught and banded as an adult in Colorado in 1976 was retrapped, alive, more than 12 years later.

rose red throat with narrow white line under bill

adult ♂

blended buffy underparts

♀

long broad tail

Rufous Hummingbird

Selasphorus rufus L 3¾" (10 cm)

This bird has the shortest breeding season and sees the longest day lengths of any hummer.

KEY FACTS

Adult male is copper-colored above with a reddish orange throat; the female has a white throat with green and red spots and buffy flanks.

+ **voice:** Hard, sharp call, *tewk*.

+ **habitat:** Summer resident of open forests and streamside groves in the West. Winters in Mexico.

+ **food:** Flower nectar, small insects, spiders.

Agile hummingbirds have almost no predators.

The Rufous Hummingbird is extremely aggressive even by hummingbird standards, with a single individual sometimes fending off dozens of other hummers to monopolize a feeder, chasing females from flowers and feeders even during breeding season. It can chase off chipmunks approaching its nest, although a tiny least chipmunk—weighing 44 grams—is about ten times heavier. Like other hummers, females feed young by regurgitation. Rufous Hummingbirds follow an elliptical migration route, heading north in spring along the coast; after the breeding season, they move south through the mountains. They also wander east more regularly than any western hummer.

rufous back

adult ♂

dark red throat

white collar

extensive rufous in tail

♀

rufous-buff flanks

Belted Kingfisher

Megaceryle alcyon L 13" (33 cm)

Unlike in most birds, female Belted Kingfishers are more colorful than males.

KEY FACTS

Stocky bird with big head, spiky crest, and large bill. Slate blue above. White below with single blue breast band in male; female has additional rusty band. Often perches over water.

+ voice: Loud rattle.

+ habitat: Common and conspicuous along streams, rivers, and coastal estuaries. Nests in streamside burrow.

+ food: Fish and other aquatic life.

Tiny feet help it shuffle through its nest burrow.

A loud dry rattle alerts us to a Belted Kingfisher flying past or hovering in midair while staring at the water and sometimes plunging in to grab prey. Kingfishers carry their catch in their bill to a perch. They pound it to stun it and break off any long spines before eating it. Both sexes excavate the tunnel-like burrow that extends several feet into the side of a riverbank, gravel pit, or similar area. It can take 3 weeks or longer to build before the female lays five to eight eggs. Chick survival is very high in these protected nests except in years of flooding.

shaggy crest

blue and rufous breast bands

single bluish breast band

white collar

Red-bellied Woodpecker

Melanerpes carolinus L 9¼" (24 cm)

This abundant woodpecker of the Southeast frequents bird feeders and backyard trees.

KEY FACTS

Familiar, zebra-striped woodpecker of the East. Male has a vibrant red crown; female has red only on her nape. White wing patch and rump visible in flight.

+ voice: Loud volley of churring notes.

+ habitat: Year-round resident of open woodlands and suburban trees.

+ food: Insects, nuts, fruits.

Hidden reddish belly feathers give them their name.

Red-bellied Woodpeckers have extended their range north and west due to a combination of factors that may include habitat change, birdfeeding, and warming trends. At feeders, they take a wider variety of foods than most woodpeckers. European Starlings often take over their nest cavities; in turn, they take nest holes from Red-cockaded Woodpeckers, sometimes injuring or even killing the endangered birds. Males incubate the three to five eggs all night; both parents share duties during the day. Except when incubating, males and females sleep alone in roost cavities. Fledglings often spend a few nights sleeping on branches before appropriating a roost cavity.

buffy forehead

red forehead

♂

barred back

♀

Yellow-bellied Sapsucker

Sphyrapicus varius L 8½" (22 cm)

Sapsuckers could more accurately be called sap-lappers; they lap sap with a brushy tongue.

KEY FACTS

Midsize woodpecker with striped face and large white wing patch. Male has red throat and forehead; female's throat is white. Juvenile is browner, lacks red.

+ voice: Nasal squeal *weeah* and catlike *meeww.*

+ habitat: Northern forests; migrates south for winter.

+ food: Eats sap and insects attracted to sap wells.

Male (left) and juvenile work the sap wells.

If forest birds have a favorite avian neighbor, it is likely to be a sapsucker. Phoebes, kinglets, warblers, and other small birds visit their sap wells. Ruby-throated Hummingbirds associate closely with them before flowers open, feeding on sap and the insects drawn to it, and cavity nesters use their nest and roost holes. The slow, arrhythmic sapsucker drumming sound is easy to recognize. When a pair of sapsuckers spends time digging a new nest cavity, they produce fewer eggs than when they reuse an old nest, yet on average they fledge more chicks from new nests, perhaps because more parasites are present in old nests.

red throat and forehead

adult ♂

white wing patch

juvenile

brownish overall with little red

Downy Woodpecker

Picoides pubescens L 6¾" (17 cm)

This diminutive woodpecker is named for the soft downy feathers on its lower back.

KEY FACTS

Black and white with a white stripe up the back and a short bill. Male has red spot on hind crown, lacking on female. Very similar Hairy Woodpecker is larger with a longer bill.

+ **voice:** Call is a sharp, high-pitched *pik!*

+ **habitat:** Year-round resident of woodlands.

+ **food:** Probes and drills into wood for insects; comes to bird feeders.

This tiny bird fluffs its feathers to conserve heat.

Charming Downies visit feeding stations for suet and seeds, and sometimes even sugar water. In winter, they're drawn to mixed flocks of chickadees and other small songbirds. Foraging in the company of other wary birds allows each to focus more on feeding and less on searching for predators. The Downy also sometimes follows a foraging Pileated Woodpecker to probe for leftovers after the huge bird finishes up. Woodpecker chicks hatch at a less developed stage than most birds, probably because carbon dioxide levels build up in deep cavities while the parents are incubating. Flying in and out to feed young raises the oxygen level.

red hind crown

♂

small bill

white back

no red

♀

Northern Flicker

Colaptes auratus L 12½" (32 cm)

Civil War soldiers from Alabama often wore feathers of the "Yellowhammer" into battle.

KEY FACTS

Eastern birds have yellow underwings; western birds are pinkish red there. Males have a black (East) or red (West) whisker mark, absent on females.

+ voice: Loud *klee-yer!*

+ habitat: Woodlands, forest edges, suburbs.

+ food: Specializes on ants, even digging into the soil for them; more berries in fall.

Woodpeckers spit out wood chips as they excavate.

The state bird of Alabama often forages on the ground, hopping or running short distances. It also feeds in trees and roosts in cavities as other woodpeckers do. Eating ants gives the flicker a strong formic acid taste. John James Audubon said its meat was "not very savory," but predators don't object—piles of flicker feathers are often found along beaches on migration pathways. In one study, raptors killed 9 percent of adults with radio tags. Despite predation, wild flickers have lived longer than 8 years. On its very first flight, a flicker chick may fly farther than 150 feet.

"Yellow-shafted" ♂

red

black "whisker"

red "whisker"

brown, barred back

"Red-shafted" ♂

Least Flycatcher

Empidonax minimus L 5¼" (13 cm)

Flycatchers belonging to the genus *Empidonax* are similar except for voice and habitat.

KEY FACTS

One of the 11 small flycatchers of the genus *Empidonax*, known as "empids" to birders. The Least is common in the East, and like other empids is olive green above with two wing bars and a white eye ring. Best identified by its song.

+ voice: Harsh *che-BEK*, usually in a series.

+ habitat: Summer resident of deciduous woods, orchards, parks.

+ food: Mostly insects.

This is one of the confusing Empidonax *flycatchers.*

The moment most birders hear *che-BEK*, they tick off Least Flycatcher and move on, often not realizing how fascinating and unique this unassuming little sprite is. Least Flycatchers nest in dense clusters rather than spreading out evenly throughout usable habitat. Each pair defends only a small territory from the others, but all aggressively chase other species from the area, perhaps why they are so seldom parasitized by cowbirds. They establish winter territories in Central America. To get the jump on the best spots, adults migrate as soon as their young are independent, not molting until they arrive in their wintering area.

worn fall adult

wing bars less obvious

conspicuous white eye ring

prominent wing bars

spring

very similar to other *Empidonax* flycatchers

Black Phoebe

Sayornis nigricans L 6¾" (17 cm)

Phoebes catch prey on the wing. They seldom walk, hop, or even pivot by foot on a perch.

KEY FACTS

Small flycatcher, mostly black with a white belly. Constantly pumps tail up and down. Stays low, often alights on streamside rocks.

+ voice: Thin, whistled *pi-tsee* or *pi-tsew.*

+ habitat: Permanent resident of open areas near water.

+ food: Mostly insects; known to catch minnows.

This phoebe often hunts from a conspicuous perch.

Calling Black Phoebes aren't as loud or insistent as their eastern relatives, but their rhythm lives up to the phoebe name. This handsome flycatcher is seldom far from water. Phoebes construct a muddy base against a structure such as a streamside boulder or open cavity and build the nest from mud and plant fibers. Natural sites are limited; nesting on manmade structures has allowed their numbers to increase. Established pairs often winter near each other and reuse nests year after year, getting a jump on breeding more quickly than those starting from scratch. Males roost on the nest rim or within a few feet of their mate.

mostly black except for white belly

juvenile

cinnamon wing bars

Eastern Phoebe

Sayornis phoebe L 7" (18 cm)

All three North American phoebes have a
distinctive habit of bobbing their tails.

KEY FACTS

Drab eastern fly-
catcher, dark brown
above with pale
underparts; no wing
bars. Constantly
pumps its tail—rapidly
down and slowly up.

+ **voice:** Male sings
a harsh *fee-bee* that
gives the species its
name.

+ **habitat:** Wooded
areas, often near
buildings. Early spring
migrant, arriving in
March in many places.

+ **food:** Flying insects.

Songbird fledglings have wide, colorful mouths.

The Eastern Phoebe has been a popular research subject for two
centuries. It was the first bird to be banded in the United States
when, in 1804, Audubon attached threaded leg bands to phoebe
nestlings and tracked them in following years. Observers have long
noted that phoebes are monogamous and seemingly devoted mates;
new DNA studies reveal that there is considerable extra-pair paternity,
particularly in the second brood each season. Banding and DNA studies confirm that some males pair with
two females. Phoebes prefer nest sites
close to overhead cover,
building taller nests to
be closer to the above
surface. They often use
phoebe nest platforms.

worn summer
adult

dark cap
and face

fresh fall

whitish belly

pale yellow
belly

Eastern Kingbird

Tyrannus tyrannus L 8½" (21 cm)

The tiny red crown, found in both sexes, is revealed only in aggressive encounters.

KEY FACTS

Midsize flycatcher with slate gray upperparts, snowy white belly, and white-tipped tail. Sits on open perches and flies out after insects.

+ voice: High-pitched sputtering notes.

+ habitat: Open areas with scattered trees for nesting. Despite its name, its range extends far to the west.

+ food: Flying insects.

Three kingbirds drive an immature Bald Eagle away.

For half the year, *Tyrannus tyrannus* lives up to its scientific name—Ben Franklin even noted when disparaging the Bald Eagle that "the little King Bird not bigger than a Sparrow attacks him boldly and drives him out of the District" (see photo). Predation on adult kingbirds is very rare, yet ironically, an American Kestrel once was observed grabbing a kingbird that was distracted because it was attacking a Red-tailed Hawk! These seasonal tyrants lead entirely different lives on their wintering grounds in the Amazon Basin. In the north, they are extremely pugnacious, territorial flycatchers. In the tropics, they become sociable vegetarians, wandering in large flocks to feed on fruits.

blackish head

white tail tip

white underparts

Western Kingbird

Tyrannus verticalis L 8¾" (22 cm)

Some winter in South Florida, and a few wander to the East Coast during migration.

KEY FACTS

Midsize flycatcher with pale gray head and back, yellow belly, and white-edged tail. Like Eastern Kingbird, chooses conspicuous perches and is highly territorial.

+ **voice:** Fussy sputtering and *kip* notes.

+ **habitat:** Summer resident of dry, open areas with scattered trees for nesting, often around farms and ranches.

+ **food:** Flying insects.

Kingbirds often feed on grasshoppers and bees.

These eye-catching birds sit on fences and other exposed perches, and conspicuously dive-bomb hawks, ravens, and other large birds. They expanded their original range as people planted trees in the Great Plains and cleared trees from heavily forested areas. Most nests are built in trees and shrubs, but many are set on such structures as utility poles, windmills, antennas, and even backyard basketball hoops, braced against the backboard. The nest is bulky, constructed of various fibers. Long strings incorporated into nests have sometimes entangled and killed nestlings; people should not set out yarns and strings longer than six inches for nesting birds.

adult

black tail with white edges

pale gray head and back

adult

yellow belly

Scissor-tailed Flycatcher

Tyrannus forficatus L 13" (33 cm)

This and a tropical relative, the Fork-tailed Flycatcher, often wander far out of range.

KEY FACTS

Astonishingly long black-and-white tail, and pinkish belly and underwings. Pale gray head and back. Juvenile has much shorter tail and is paler overall.

+ voice: Fussy sputtering and *pup* notes.

+ habitat: Summer resident of grasslands and prairies with scattered trees.

+ food: Insects, especially grasshoppers, crickets, and beetles.

This bird is beautiful in both form and color.

A sitting Scissor-tailed Flycatcher is splendid enough. Sallying out to capture a flying insect, it exposes salmon pink wing linings and, as it hovers or turns abruptly, scissors its spectacular tail. In spring, males wheel and dip in ethereal courtship flights that leave even the most gushing superlatives wanting. Oklahoma's state bird was well chosen—its small breeding range is centered on the state. From spring through fall, it's hard to miss these extraordinary birds sitting conspicuously on fences and other obvious perches.

adult ♂

reddish pink "armpit"

juvenile

much shorter tail

adult ♂

orange-buff belly

long, forked tail

Loggerhead Shrike

Lanius ludovicianus L 9" (23 cm)

Nicknamed "butcher birds," shrikes are predatory songbirds that impale prey on thorns.

KEY FACTS

Small gray, black, and white songbird with the hooked bill of a predator. Similar in coloration to Northern Mockingbird (page 141), but note the shrike's black mask.

+ voice: Harsh *shack-shack*.

+ habitat: Open country with scattered bushes. Declining in most areas.

+ food: Large insects, lizards, mice, and birds.

This small predator hunts from a conspicuous perch.

Lacking talons, the shrike uses its sharp bill to grab a victim, and dispatches it with a bite to the nape. Using its bill again, the shrike carries off the carcass and impales it on a barb or thorn; the food may be eaten on the spot or saved for later. This 2-ounce songbird has a high metabolic rate and must eat frequently. When hunting is good, food accumulates to be used when hunting is poor. Mockingbirds, caracaras, and Burrowing Owls sometimes raid the shrike's food stores, as do neighboring shrikes. Mated birds that don't migrate seem to stay close to each other year-round.

hooked bill

black mask

adults

white wing patch

Red-eyed Vireo

Vireo olivaceus L 6" (15 cm)

This unassuming bird is among the top ten
most abundant land birds in North America.

KEY FACTS

Small, olive forest bird
with striped face and
gray crown bordered
with black; the red eye
is hard to see.

+ voice: Series of
singsong phrases
(*Here I am, over here,
see me, where are
you?*).

+ habitat: Summer
resident of mature
woodlands.

+ food: Insects, par-
ticularly caterpillars.

Vireo nests are suspended from a forked branch.

These often-overlooked birds sing
what seems to be an endless
monologue of robin-like phrases
of two or three notes, pausing
between each phrase as
if for dramatic emphasis.
They continue long after most songsters
have quit for the day. The song is pleasing and
the birds ubiquitous in summer, found in virtually every forest
and small woodlot, yet many people live out their lives without
ever knowing they exist. They quickly notice jays, cats, and other
predators and draw the attention of other birds by scolding with
querulous, harsh mews. Their compact cup nest suspended
from a forked branch usually holds three to four young.

olive
upperparts

striped
head

red eye hard
to see

Blue Jay

Cyanocitta cristata L 11" (28 cm)

Jays bury more acorns than they eat and have helped replant forests after glaciers melted.

KEY FACTS

Easily recognized by its blue color, crest (sometimes flattened), black necklace, and bold white spots. Male and female look alike.

+ voice: Wonderfully diverse, including a piercing *jay, jay, jay.*

+ habitat: Mixed woodlands, parks, suburbs.

+ food: Omnivorous. Nuts, fruit, insects; visits bird feeders.

Mated pairs stay together through winter.

The provincial bird of Prince Edward Island, one of the most handsome, charismatic birds of America, attracts our notice with its loud squawks and colorful plumage. Feathers have no blue pigments; specialized cells in the barbs scatter light to produce blue. Adult jays feed primarily on seeds and fruits, along with insects during summer. Their four to five chicks require a high-protein diet, and so are fed insects supplemented with nestling birds of other species. When jays investigate a nest, many songbirds join together to drive them away. The rest of the time, jays are valued neighborhood watch guards, alerting everyone to danger.

bluish crest

black throat band

extensive white in wings and tail

Steller's Jay

Cyanocitta stelleri L 11½" (29 cm)

Corvids—crows, ravens, magpies, and jays—are among the most intelligent birds in the world.

KEY FACTS

Dark blue and black plumage, and long, shaggy crest. Pale forehead streaks are blue on most, white on birds in southern Rockies.

+ voice: Piercing *shack, shack, shack.*

+ habitat: Forest resident found from sea level to high elevations; frequents picnic areas looking for handouts.

+ food: Omnivorous. Nuts, fruit, insects; visits bird feeders.

This inquisitive jay often mooches from people.

long black crest

pale forehead streaks

no white in wings

America's darkest jay, the provincial bird of British Columbia, is much more likely to visit campers for handouts than the Blue Jay, but is not as bold as its relatives, the Gray Jay and Clark's Nutcracker. It raids caches from them and from Acorn Woodpeckers. It has complex social interactions and vocalizations, including harsh notes, imitations of other species, and a quiet warbling song. Steller's Jays from a wide area join together to mob predators. In one case, a small group struck a perched Cooper's Hawk on the back, knocking it to the ground. Mated pairs remain together for life and are socially dominant over unmated individuals.

Western Scrub-Jay

Aphelocoma californica L 11" (28 cm)

This jay has a curious habit of bobbing in several directions when it alights on a perch.

KEY FACTS

Familiar "blue jay" of the West—unlike Blue Jay of the East (page 112)—has no crest or prominent white spots. Coastal scrub-jays are conspicuous; interior birds are shy and quiet.

+ **voice:** Harsh, up-slurred *jay?* or *jreee?*

+ **habitat:** Year-round resident of chaparral, open woods, and backyards.

+ **food:** Omnivorous. Nuts, fruit, insects; visits bird feeders.

These jays can survive in very arid environments.

Unlike its close relative, the Florida Scrub-Jay, this species is extremely vocal, sometimes following people while making harsh calls that warn other birds away. It preys on eggs and young of other species more often than Blue Jays do, and also hunts a variety of rodents and lizards. Scrub-jays wedge acorns and other hard nuts into tree crevices to hammer them open. Like other jays, they place a leaf or two over each cached food item they hide on the ground. Those in the habit of raiding other birds' caches are the most suspicious about being observed while hiding their own stores. Birds living in the interior West may be a separate species.

gray-brown back

partial blue breast band

long tail

||||||||||||||||||||||||||||||||||||||||||||||||||||||||||||||||||||||||||||||||||||||||||||||||||||

Black-billed Magpie

Pica hudsonia L 19" (48 cm)

Magpie nests are huge masses of sticks with two entrances to an inner nest cup.

Magpies often follow predators to share the meal.

KEY FACTS

Striking black-and-white plumage; long, glossy tail and bluish wings. Magpies sometimes gather in large flocks. White wing patches visible in flight.

+ voice: Various harsh calls and raspy chatter.

+ habitat: Year-round resident of western rangelands and foothills.

+ food: Omnivorous. Fruit, grain, insects, small animals, carrion.

Magpies raided Lewis and Clark's tents and stole food from the expedition. The explorers shipped four magpies to President Jefferson from Fort Mandan in 1845; only one survived the journey. This spectacular corvid scavenges on dead animals and sits on the backs of large mammals, picking off and eating ticks. It also may kill or injure small farm animals, and sits on large animals pecking flesh from open sores and wounds. Some states once offered a bounty on them. This extremely intelligent species can recognize its reflection in a mirror. It was considered the same species as the European magpie until 2000.

large white wing patch

black-and-white plumage

very long tail

||||||||||||||||||||||||||||||||||||||||||||||||||||||||||||||||||||||||||||||||||||||||||

American Crow

Corvus brachyrhynchos L 17½" (45 cm)

Crows remember individual people perceived as threats and teach other crows to fear them.

KEY FACTS

All-black plumage. Larger than grackles and blackbirds; smaller than ravens. In the South, where the Fish Crow (*uh-uhh* call) also lives, the two species are best told apart by voice.

+ **voice:** Familiar call is harsh *caw, caw, caw.*

+ **habitat:** Various, especially open areas with scattered trees.

+ **food:** Grain, insects, small animals, carrion, garbage.

Young crows have red mouths, some for over a year.

Henry Ward Beecher wrote, "If men had wings and bore black feathers, few of them would be clever enough to be crows." These fascinating birds maintain family and neighborhood ties that last generations. When West Nile virus decimated crows in Ithaca, New York, where many crows are individually marked for study, one widowed female adopted her dead neighbors' orphans; the orphans later helped her raise her own young. In fall and winter, crows join enormous flocks. When one detects a Great Horned Owl, it makes a loud "assembly call," bringing in dozens more that noisily and aggressively harass the owl. Many crows fly over highways searching for fresh roadkill.

all-black plumage

slightly rounded tail tip

often in flocks

Common Raven

Corvus corax L 24" (61 cm)

The largest songbird in the world is the provincial bird of Yukon Territory.

KEY FACTS

All-black plumage and heavy bill. Often soars, unlike much smaller crows. Ravens have shaggy neck feathers and a wedge-shaped tail.

+ voice: Common calls are a croaking *kraaah* and a hollow *brooonk.*

+ habitat: Widespread in the West; uncommon in the East.

+ food: Grain, insects, small animals, carrion, garbage.

Ravens are opportunistic, smart, and devoted mates.

Many authorities consider the raven one of the most intelligent animals on earth. It features as a benevolent trickster in native folklore of the Pacific Northwest. Its close associations with people, scavenging habits, and dark plumage contributed to its role as Poe's "ghastly grim and ancient" bird. Ravens have been associated with the Tower of London for centuries; they may have arrived to scavenge on executed corpses or victims of the 1666 Great Fire of London. A flock of seven captive ravens is still maintained in the Tower because of a superstition that if they disappeared, "the Crown will fall and Britain with it."

size of large hawk

heavy bill

all-black, shaggy plumage

wedge-shaped tail tip

Horned Lark

Eremophila alpestris L 7¼" (18 cm)

America's only native member of the lark family is not related to meadowlarks.

KEY FACTS

Small songbird with black mask and forehead that extends upward as two "horns." Pinkish brown above with variable amount of yellow on face and a black bib. Forms large flocks in winter.

+ voice: Song is series of tinkling notes.

+ habitat: Open country, prefers bare ground.

+ food: Seeds; feeds insects to its young.

Tiny feathers form "horns" that can be hard to see.

A lovely yet unobtrusive song rings in the skies over farm country, airports, and other open areas. America's closest relative of Percy Bysshe Shelley's blithe spirit has a spectacular sky dance, flying straight up as high as 800 feet where it breaks into its high-pitched, tinkling flight song while circling for many seconds. When finished, it drops headlong back to earth, wings closed until the last moment. Horned Larks blend in so well that they're usually overlooked on the plowed fields and rocky bare ground where they live, even when they gather in large flocks, but a close-up view through a scope or binoculars is well worth the effort.

white underwings

black "horns"

black tail

black "bib"

variably yellow (to white) on throat and face

||||||||||||||||||||||||||||||||||||||||||||||||||||||||||||||||||||||||||||||||||||||

Purple Martin

Progne subis L 8" (20 cm)

North America's largest swallow spends winters in the Amazon Basin.

KEY FACTS

Male is dark purple overall; female and juvenile are dingy grayish white below. In flight, rapid flapping alternates with short glides.

+ voice: Loud, rich gurgling and whistles.

+ habitat: Open fields. Eastern birds nest exclusively in human-provided gourds and apartment houses. Migrates to South America.

+ food: Flying insects.

Martins have wide mouths for flycatching.

When a Purple Martin's tiny bill opens, revealing a truly capacious mouth, it looks as if the face split wide open. During nesting season, they dine on flying insects and make rich, liquid burblings and twitterings. Martins in the West and in Mexico nest almost exclusively in natural cavities. East of the Rockies, there is a dramatic switch. There, virtually all martins nest in birdhouses and hollowed-out gourds, and there are very few records of them nesting in natural cavities during the entire 20th century. Because so many martin houses are set out to accommodate several pairs, most martins in the East nest colonially.

eastern ♀

pale belly

adult ♂

dark purple overall

much paler than eastern female

western ♀

Tree Swallow

Tachycineta bicolor L 5¾" (15 cm)

The Tree Swallow can digest berries when temperatures are too low for flying insects.

KEY FACTS

Adult has dark blue (or greenish), glossy upperparts and pure white belly. Tail is slightly forked. Juvenile is grayish brown above with diffuse band across chest.

+ voice: Liquid twittering and chirping.

+ habitat: Needs open habitat and prefers to nest near water.

+ food: Flying insects; small fruits in winter.

Hungry and demanding nestlings keep parents busy.

The first swallow to arrive north in spring, it nests in birdhouses, natural cavities, and woodpecker holes. It builds a nest of grasses and straws lined with white feathers. Nesting success is directly correlated with the number of feathers, which insulate eggs and chicks from extreme temperatures. Prairie grouse get into skirmishes while displaying—when they leave, Tree Swallows pick up the feathers the big birds lose. People can set out clean poultry feathers to help swallows. They complete breeding in early July and form enormous flocks. By late fall in the South, these flocks can number in the hundreds of thousands.

adult

1st spring ♀

dark brown upperparts

adult

dark blue upperparts

white below

spring adult

||||||||||||||||||||||||||||||||||||||||||||||||||||||||||||||||||||||||||||||||||

Cliff Swallow

Petrochelidon pyrrhonota L 5½" (14 cm)

Cliff and Barn Swallows gather at puddles to pick up mouth-size plops of mud for nests.

KEY FACTS

Compact body with short, square-tipped tail and buffy rump patch. Adult has white forehead and chestnut cheeks and throat. Juvenile is browner with less evident head pattern.

+ voice: Squeaking twitters and grating notes.

+ habitat: Open areas with overhanging cliffs or structures to attach their mud nests to.

+ food: Flying insects.

In colonies, mud nests are clustered together.

Cliff Swallows build their adobe houses in large groups; a Wisconsin farmer once counted 2,015 nests on his barn. Females may start laying eggs before the roof of the nest is completed. The swallows nesting at San Juan Capistrano used to return every March when weather was warm enough to sustain flying insects. For people who carefully kept their eyes averted until March 19, the swallows indeed appeared on St. Joseph's feast day. They disappeared from the mission in recent years, probably in response to major landscape changes. Historically, the population spread from the West across the plains to the East, and numbers overall are fairly stable.

Southwest birds have rufous forehead

white forehead

buffy rump

reddish brown throat

Barn Swallow

Hirundo rustica L 6¾" (17 cm)

Barn Swallows build a large cup nest;
Cliff Swallows a globe with an entrance hole.

KEY FACTS

Cobalt blue above and buffy below with a chestnut throat. Long, graceful tail of adult is unlike any other North American swallow; juvenile's tail is shorter.

+ voice: Series of scratchy, warbling notes and grating rattles.

+ habitat: Summer resident of open areas; shelters its mud nest on man-made structure.

+ food: Flying insects.

Parents sometimes feed fledglings in flight.

Aristotle may have been thinking of this bird when he said, "One swallow does not a summer make." The most widespread swallow in the world nests under bridges and eaves, and in culverts, barns, and outbuildings. In Minnesota, a pair nested for several years in the lumber section of a home improvement store, hovering in front of the sensor to trigger the door to open so it could pass in and out. New World Barn Swallows once bred only in North America, spending the winter in South America as nonbreeders. In 1980, six pairs nested in Argentina. Now, a South American breeding population is well established.

long, forked tail

paler below than adult

juvenile

shorter tail

bluish above

||||||||||||||||||||||||||||||||||||||||||||||||||||||||||||||||||||||||||||||||||||||||||||||

Black-capped Chickadee

Poecile atricapillus L 5¼" (13 cm)

In late summer, sibling chickadees disperse, each joining a different winter flock.

KEY FACTS

Small, energetic songbird with oversize head patterned in black and white. The very similar Carolina Chickadee is found in the South.

+ voice: Call is a slow *chick-a-dee-dee-dee.*

+ habitat: Common resident of woodlands, wooded edges, suburbs, towns.

+ food: Insects, seeds, and berries; often visits bird feeders.

Sunflower seeds are relished, in gardens or feeders.

The chickadee's inquisitive ways and acrobatic habits endear it to people, from hunters in deer stands to the housebound at their window. Chickadees are often the first birds to discover a new feeder and sometimes take seeds or mealworms from the hand. They nest in cavities they excavate themselves or in woodpecker holes and birdhouses.

worn summer

reduced white on wings

flanks mostly white

white cheeks

white edges

buffy

fresh fall

They prefer birdhouses filled with wood shavings for them to remove. Their song is a pure whistled *hey, sweetie.* When scolding, the more *dee* notes in their *chick-a-dee-dee-dee* call, the more dangerous the threat. Chickadees maintain strict hierarchies in winter flocks, which often include other songbirds.

Tufted Titmouse

Baeolophus bicolor L 6¼" (16 cm)

This handsome, crested relative of the chickadee sings year-round.

KEY FACTS

Larger than a chicka-dee, with an obvious crest and a small black forehead. Juvenile lacks the black forehead.

+ voice: Song is whistled *peter-peter-peter*; call is chickadee-like *tsicka-dee-dee.*

+ habitat: Permanent resident of eastern forests and suburbs.

+ food: Insects, seeds, and berries; visits bird feeders.

Landing gear out, this bird is ready to alight.

Titmice nest and roost in cavities and birdhouses. To line their nests, they find soft fur, even plucking tail hairs from road-killed squirrels and sleeping raccoons. Some titmouse young disperse to other areas in late summer as chickadees do, but some remain with their parents all winter, and a few help their parents raise the following year's brood. Pairs are territorial year-round. Winter groups include a mated pair, their young, and young that dispersed from other areas; sometimes other small songbirds join the group. The Black-crested Titmouse of Texas and Mexico was until recently considered the same species.

lacks black forehead

juvenile

tall crest

large black eye

black forehead

adult

||||||||||||||||||||||||||||||||||||||||||||||||||||||||||||||||||||||||||||||||||||

Bushtit

Psaltriparus minimus L 4½" (11 cm)

Bushtits, sociable little birds, are famous for their enormous, dangling pouch nest.

KEY FACTS

Tiny gray-brown songbird with a plump body and long tail. Male has dark eyes; female has pale eyes. Flocks of 10 to 40 birds forage together. Builds unusual hanging nest.

✛ **voice:** Flocks twitter constantly.

✛ **habitat:** Permanent resident of western thickets, scrublands, and backyards.

✛ **food:** Small insects and spiders.

A Bushtit peeks out from its well-constructed nest.

Active little Bushtits have two color variations that were once classified as different species. Researchers discovered that "black-eared" birds are juvenile males that are most common in Mexico, not a separate species. Bushtits huddle together at night or when resting by day, sometimes even when it isn't cold. Mated pairs take about a month to build their nest, which is huge for a bird weighing little more than two pennies. The nest maintains a more constant internal temperature, allowing the birds to spend more time foraging instead of incubating. In some places, the parents get help raising their young from unattached Bushtits.

♀
pale eye
very small
mostly gray
long tail
usually in a flock
dark eye
♂

Red-breasted Nuthatch

Sitta canadensis L 4½" (11 cm)

This is the only nuthatch that migrates regularly, sometimes as early as July.

KEY FACTS

Chickadee-size song-bird with black-and-white head stripes and cinnamon underparts. Moves with a jerky, toylike action. Often very tame.

+ **voice:** Series of nasal calls *ehhnk ehhnk.*

+ **habitat:** Coniferous woods and mountain forests; some wander south in winter.

+ **food:** Insects, spiders, seeds.

A bold nuthatch scolds a sleepy Barred Owl.

These handsome birds of fir and spruce forests excavate their own nest cavities. The female chooses a dead branch and starts digging. As she enlarges the interior nesting space, she tosses sawdust out the entrance—it accumulates at the base of the tree. Both the male and female smear globules of sticky conifer resin around the entrance, which probably deters predators. The parents dive into the nest with precision to avoid touching the resin, though at least one female was found dead, firmly stuck. When conifer seeds are scarce, large numbers head south of their breeding range in "irruptions."

gray cap

paler below than male

♀

♂

cinnamon underparts

prominent white supercilium

||||||||||||||||||||||||||||||||||||||||||||||||||||||||||||||||||||||||||||||||||||||||||||||||||||||||||||

White-breasted Nuthatch

Sitta carolinensis L 5¾" (15 cm)

Nuthatches forage in bark crevices while upside down, finding insects that other birds miss.

Short tails characterize nuthatches in flight.

KEY FACTS

Larger than a chicka-dee, with a stocky body, short tail, and dagger-like bill. Male has black cap, blue-gray upperparts; female has gray cap, dull gray back.

+ **voice:** Low-pitched nasal *yank*; higher-pitched in the West.

+ **habitat:** Forests, wooded suburbs, parks.

+ **food:** Insects, spiders, seeds.

America's only nuthatch associated with deciduous forests doesn't have stiff tail feathers to brace itself against a trunk. Unlike woodpeckers and creepers, nuthatches have very short tails and a long back claw. The name "nuthatch" comes from the habit of wedging seeds and nuts into crevices of bark and hacking them open with the bill. White-breasted Nuthatches are territorial year-round, though they may leave their territory to visit feeders when food is scarce. After a pair is established, they remain together as long as they both live. They don't try to nest more than once a year, even if their first attempt fails.

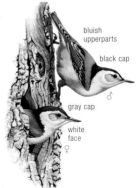

bluish upperparts

black cap

gray cap

white face

♂

♀

Brown Creeper

Certhia americana L 5¼" (13 cm)

Creepers have forward-facing eyes that can examine bark closely with binocular vision.

KEY FACTS

Brown upperparts are mottled with pale streaks and dots that closely match the furrowed bark it clings to. Note white eyebrow; long, spiky tail; and cinnamon rump.

+ voice: Call is a very high-pitched *seee.*

+ habitat: Mature forests. Birds in Southwest mountains may be a separate species.

+ food: Insects, spiders.

Creepers build their nest behind flaps of bark.

The Brown Creeper's quiet habits, high-frequency songs and calls, and cryptic plumage make it one of the most inconspicuous of songbirds. To feed, a creeper begins at the base of a tree trunk and climbs upward, sometimes following a straight path, sometimes spiraling, until it reaches obstructing limbs or the top, when it drops like a leaf to the base of a nearby tree to start the process again. When it detects an insect or spider in the bark, it snaps it up with its delicate, curved bill. Creeper nests are so well hidden behind loose flaps of bark on decaying trunks that ornithologists didn't discover one until 1879.

thin, curved bill

streaked and spotted upperparts

long, spiky tail

Cactus Wren

Campylorhynchus brunneicapillus L 8½" (22 cm)

Cactus Wrens get all the water they need from their food, only rarely drinking at birdbaths.

KEY FACTS

Very large wren with brown crown, streaked back, heavily barred wings and tail, broad white eyebrow. Breast is densely spotted with black, less so in coastal California.

+ voice: Rapid series of harsh notes *cha cha cha*, heard all year.

+ habitat: Conspicuous denizens of cactus country, arid scrub, and desert suburbs.

+ food: Mostly insects and spiders.

These wrens have no problem perching on cacti.

The well-chosen state bird of Arizona is perfectly adapted to extreme desert conditions. Males sing a rapid series of grating notes from conspicuous perches on and off all day, even in the heat of the afternoon. Females more rarely sing, their notes softer and higher pitched. Each pair builds several bulky, globular nests within spiny plants, and uses them for roosting year-round. They often take dust baths before retiring at dusk. Parents continue to feed their chicks for weeks after they fledge. Young birds may start nest-building activities just 12 days after fledging, using the structures for nighttime roosting.

ball-shaped nest

densely spotted

banded tail

very large wren

Carolina Wren

Thryothorus ludovicianus L 5½" (14 cm)

The tail of South Carolina's state bird is usually cocked up, but it's lowered when singing.

KEY FACTS

Chunky, rufous-brown wren with buffy underparts, bold white eyebrow, and long twitchy tail. Pairs stay together year-round.

+ **voice:** Vocal throughout the year; song is a loud, rolling *tea-kettle tea-kettle tea-kettle*.

+ **habitat:** Eastern woods, brushy ravines, and backyards.

+ **food:** Pokes into crevices for insects and spiders.

This handsome wren is carrying nesting material.

This backyard bird nests in crevices and cavities including birdhouses, glove compartments in abandoned cars, old shoes, mailboxes, flowerpots, and pockets of coats on clotheslines. They are much more often heard than seen. Their song is loud, but pleasing and seldom obtrusive. Many tropical wrens sing complex duets, but the Carolina Wrens' duets are more primitive. The male sings the familiar song, and occasionally the female chimes in with a buzzy phrase. Although nonmigratory, Carolina Wrens wander, and the species expanded its range northward in the 20th century, due in part to milder winters, reforestation of some areas, and bird feeders.

bold white supercilium

rufous-brown upperparts

buffy underparts

||||||||||||||||||||||||||||||||||||||||||||||||||||||||||||||||||||||||||||||||||||

House Wren

Troglodytes aedon L 4¾" (12 cm)

House Wrens may add spider egg cases to nest material; hatching spiders eat nest parasites.

KEY FACTS

Small, nondescript wren with finely barred brown plumage and medium-length tail, often held cocked up. Very assertive; will drive larger birds from nest hole it wants.

+ **voice:** Cascade of whistled notes.

+ **habitat:** Summer resident across North America; found in open woods, thickets, and backyards.

+ **food:** Mostly insects and spiders.

House Wrens nest in all kinds of boxes and holes.

House Wrens sing a cheerful, bubbly song virtually all day, are easily attracted to birdhouses, and eat a wide variety of insect pests, making them treasured backyard birds for all but those maintaining bluebird houses. House Wrens often peck or remove eggs and young nestlings of other nearby cavity nesters. Yet, unexpectedly, they've also been recorded feeding chicks of other species. While his mate was incubating, one male started feeding nestling flickers in the same tree. Several days later, when his own chicks hatched, he continued feeding the large woodpeckers as well as his own tiny young. House Wrens often take new mates for second broods within a season.

indistinct head pattern

nests in hole

Golden-crowned Kinglet

Regulus satrapa L 4" (10 cm)

Songs and calls of this tiny species are too high pitched for many people to hear.

KEY FACTS

Tiny and very active. Male has black crown with yellow and red center; female's crown has a yellow center. Pale gray below.

+ voice: Call is a very high pitched, often overlooked *tsii tsii tsii.*

+ habitat: Breeds in coniferous forest. Some birds move south in winter and are found in various wooded habitats.

+ food: Insects, spiders.

This male is fully displaying his crown.

A flame-colored crown is appropriate on a bird exuding as much heat as a Golden-crowned Kinglet. While the female is actively incubating, her eight to nine tiny eggs are kept at about 104°F. If that seems improbably warm, beneath her feathers, her body is about 111°. Many weigh less than a nickel, yet can survive winter in northern areas until temperatures drop to about -40°. Even in the dead of winter, they fuel their metabolic furnace on insects. In spring, they construct a tiny, thick-walled nest that will stretch out as their chicks grow. They nest twice a year but don't reuse their nests.

orange crown patch bordered by yellow and black

♂

white supercilium

yellow crown patch

♀

tiny size

Ruby-crowned Kinglet

Regulus calendula L 4¼" (11 cm)

This kinglet's loud, rich warble includes notes sounding like *liberty, liberty, liberty.*

KEY FACTS

Tiny and very active. Slightly larger than Golden-crowned with more uniform olive plumage. Blank face with white eye ring. Male has red crown.

+ voice: Call is a husky, scolding *je-dit.*

+ habitat: Breeds in coniferous forest; migrates south and spends the winter in various wooded and brushy habitats.

+ food: Insects, spiders.

This male's red crown is partially exposed.

Kinglets flit about while feeding, seldom remaining in any spot for more than a second or two, and constantly flicking their wings. The male Ruby-crown's song is exceptionally loud for his size; he often sings a muted version immediately before returning to the nest when eggs or chicks are present. Ruby-crowns produce the largest clutch of tiny North American songbirds—up to 12 eggs, which must be layered in the narrow nest. Females probably use their legs as well as the patch of bare skin on their belly (the brood patch) to keep all the eggs evenly warm. In winter, both kinglets join flocks, often with chickadees.

broken eye ring

red crown patch usually hidden

active, often flicks wings

♂

Blue-gray Gnatcatcher

Polioptila caerulea L 4¼" (11 cm)

Other gnatcatchers are nonmigratory, but the Blue-gray retreats south in winter.

KEY FACTS

Small, slender song-bird; gray above and white below, with a long, twitchy tail. Breeding male has a thin black eyebrow, absent in winter.

+ **voice:** Call is a querulous *speeeee.*

+ **habitat:** Wide-spread in woodland settings; winters in southern states.

+ **food:** Insects, spiders.

Nests of spider silk can stretch as chicks grow.

People seldom notice this tiny bird with its soft, high-frequency sounds, unless they're specifically looking for it. But a glimpse at a gnatcatcher richly rewards a search through the foliage. It is pugnacious, often attacking predators such as Cooper's Hawk and Northern Pygmy-Owl, diving at and hovering around their head and persistently following them. The tiny nest is built with lichens and spider silk, expanding as the four to six nestlings grow. This is the tiniest species regularly parasitized by Brown-headed Cowbirds. Cowbird eggs are too large for gnatcatchers to pierce or toss out, and survival of gnatcatcher chicks in a parasitized nest is very low.

prominent eye ring

long, twitchy tail

♀

thin black line over eye

pure white outer tail feathers

breeding ♂

American Dipper

Cinclus mexicanus L 7½" (19 cm)

The "water ouzel" lives along rushing western streams with clear, unpolluted water.

KEY FACTS

Aquatic songbird with a heavy, rounded body. Adult is uniformly sooty gray with a short tail and dark bill. Juvenile is paler overall with faint barring on underparts.

+ **voice:** Ringing, wren-like song of musical notes; audible over the din of rushing water.

+ **habitat:** Mountain streams in the West; lower in winter.

+ **food:** Aquatic insects and their larvae.

Dippers eat aquatic insects and small fish.

The drab dipper elicits gasps of admiration from the graceful way it bobs up and down while standing on wet rocks in an icy river and then jumps in, walking underwater to search for food. Dippers nest on horizontal ledges or crevices, under overhanging dirt banks, or other noisy streamside sites, sometimes behind waterfalls. About half the nests are constantly sprayed, but there isn't a clear advantage for either wet or dry nests. To survive freezing temperatures in and out of frigid waters, dippers have a low metabolic rate, high oxygen-carrying capacity in their blood, and very thick plumage.

yellowish bill

juvenile

chunky body and gray overall

paler underparts

often bobs body

will swim underwater to obtain food

Eastern Bluebird

Sialia sialis L 7" (18 cm)

Missouri and New York's state bird is beloved for its beautiful plumage and soft, lovely song.

KEY FACTS

Male has flashy blue upperparts; female's plumage is more subdued. Rusty color on breast wraps up and forms a partial collar; belly is white. Juvenile is heavily spotted.

+ **voice:** Rich, musical warble *chur chur-lee chur-lee.*

+ **habitat:** Open, rural areas. Most birds do not migrate.

+ **food:** Insects from spring to fall; small fruits in winter.

Both parents feed nestlings throughout the day.

Many people buy birdhouses in hopes of attracting bluebirds. But these red, white, and blue birds that inspired the "bluebird of happiness" are not backyard birds—they live in expansive fields, pastures, and orchards. Providing a few perches from which they can scan the ground for insects increases the likelihood of attracting them. Nesting males and females can be ferocious toward competitors. When pairs succeed in raising young, they often pair together the following year. Bluebirds increased when settlers cleared forests, and declined in the mid-20th century. Restoration projects helped bring back their numbers.

paler than male ♀

spotted

juvenile

rich rufous below

white belly ♂

Mountain Bluebird

Sialia currucoides L 7¼" (18 cm)

The state bird of Idaho and Nevada sometimes wanders east as far as Long Island in winter.

KEY FACTS

Male is sky blue above, paler blue below. Female is brownish gray overall with pale blue wings and tail, and flanks tinged with chestnut. Juvenile is spotted below.

+ voice: Warbled *tru-lee*; call is a thin *few*.

+ habitat: Western montane habitats; migrates to lowlands in winter.

+ food: Insects, small fruits.

Bluebirds nest in cavities as well as birdhouses.

The Mountain Bluebird eats fewer berries and other plant food than other thrushes, and searches the ground for insect prey by hovering, more in the manner of a kestrel than a bluebird. As soon as a female accepts a mate, he starts following her closely wherever she goes until the chicks hatch. He does this "mate guarding" as she gathers nesting materials, even though he doesn't contribute to nest building. Most of the day, Mountain Bluebirds are quieter than other bluebirds, singing their soft, rich song most frequently before first light. Their alternate song, a soft, repetitious warble, can be sung at any time, but is often overlooked.

often hovers above prey

♀ brownish gray and pale bluish plumage

♂

sky blue plumage

Hermit Thrush

Catharus guttatus L 6¾" (17 cm)

Vermont's state bird only rarely sings its beautiful carol in winter or during migration.

KEY FACTS

Medium-size thrush—smaller than a robin—with a bright rufous tail. Upperparts are rich brown and the breast is buffy and spotted. Western birds are grayer brown. Juvenile has spotted upperparts.

+ **voice:** Series of clear, flutelike notes; call is a harsh *chup*.

+ **habitat:** Boreal and high-elevation forests; winters farther south.

+ **food:** Mostly insects and small fruits.

Thrushes have breast spots and a robin-like shape.

The Hermit Thrush's down-to-earth feeding habits, nest placement on the ground or low in trees, and understated earth-tone plumage belie the heavenly images conjured by its song. Rather than a simple larynx, birds have an intricately designed syrinx where the trachea splits into the bronchial tubes, allowing them to create multiple sounds simultaneously.

pale spots on back and head

juvenile

boldly spotted breast

contrasting reddish tail

Ornithologists have chosen religious mnemonics to describe this thrush's ethereal song, such as *oh, holy holy, ah, purity purity, eeh, sweetly sweetly*. In spring, Hermit Thrushes arrive weeks earlier than other forest thrushes. This is the only forest thrush whose population has increased or remained stable over the past two decades.

||||||||||||||||||||||||||||||||||||||||||||||||||||||||||||||||||||||||||||||||

American Robin

Turdus migratorius L 10" (25 cm)

This state bird of three states spends fall and winter in sociable flocks feeding on fruits.

KEY FACTS

Fairly large, potbellied songbird with brick red breast, gray back, blackish head. Juvenile has paler breast with spots.

+ **voice:** Loud, musical song *cheerily cheer-up cheerio.*

+ **habitat:** Common in many places, including lawns and parks, also wilder locations. Forms flocks in winter.

+ **food:** Earthworms and insects; more berries and fruit in winter.

A chick takes 24 hours to pip its shell and hatch.

Named by homesick settlers for the familiar European "robin redbreast" (unrelated to thrushes), our robin is a thrush belonging to the same genus as the Eurasian Blackbird. Some robins spend all or part of the winter in northern areas where fruits remain. When temperatures average about 37°, males return to territories, sing, and switch to high-protein food. Females arrive a few days later. Both parents feed nestlings. When the young fledge, the female renests as the male attends the fledglings. When the new clutch hatches, the older chicks are ready to be on their own.

white spot

juvenile

spotted underparts

yellow bill

brick red underparts

♂

Gray Catbird

Dumetella carolinensis L 8½" (22 cm)

Catbirds are best known for their mewing calls; males sing a long jumble of mixed sounds.

KEY FACTS

Slender, dark gray bird with a long tail, black cap, and chestnut undertail. Stays in underbrush and makes short, low flights.

+ voice: Mix of melodious and squeaky notes; call is a catlike *mew.*

+ habitat: Thickets, brushy forest, vine tangles; withdraws to the south in winter.

+ food: Insects; more small fruits in winter.

The rusty "crissum" is often hard to see.

Catbirds are easier to hear than see. They skulk in dense underbrush, and when one emerges to sing on an exposed branch, he often perches with his tail depressed, his slender, almost branch-like form keeping him hidden in full view. Often when we think catbirds must be nesting nearby because of their sounds, it isn't until after leaves fall that the nest is revealed in a conspicuous shrub. Catbirds eject cowbird eggs. They often destroy eggs and nestlings of other nearby birds, yet have been reported caring for orphaned young of other species. They may visit feeders for fruits, jelly, or mealworms. At least one wild catbird lived more than 17 years.

blackish cap

chestnut undertail coverts

Northern Mockingbird

Mimus polyglottos L 10" (25 cm)

Mockingbirds sing a long string of repeated phrases, adding new ones throughout their life.

KEY FACTS

Slender, gray-and-white songbird with a long tail and yellow eyes. Large patches of white in the wings and outer tail feathers. Juvenile has spotted breast and dark eyes.

+ voice: Mix of original and mimicked phrases, each repeated two to six times.

+ habitat: Thickets and brushy areas; most birds are resident.

+ food: Insects; more small fruits in winter.

In flight, mockingbirds are very conspicuous.

The state bird of five states was one of Thomas Jefferson's favorite birds; he kept one as a pet in the White House. A male's repertoire may consist of more than 150 distinct song types; the number increases with age. Imitations may include mechanical sounds, vocalizations of other animals, and human voices, laughter, and screams. Unattached males may sing throughout the night. Pairs rear as many as four broods in a season. On the ground, mockingbirds frequently raise their wings to flash the white wing patches, perhaps to startle insects or predators or serve as a territorial display.

long tail

white wing patch

spotted underparts

juvenile

white outer tail feathers

Brown Thrasher

Toxostoma rufum L 11½" (29 cm)

Thrashers are ground-foraging mimids; this species sings most imitations twice.

KEY FACTS

Rich rufous upper-parts, long tail, two whitish wing bars, and yellow eyes. Underparts are extensively streaked. Skulks in the underbrush.

+ **voice:** Series of melodious phrases, each repeated two or three times.

+ **habitat:** Eastern species favors dense cover; winters in the South.

+ **food:** Insects; more small fruits in winter.

Yellow eyes distinguish thrashers from thrushes.

Georgia's state bird may have even more song types than mockingbirds—in 1981, a Brown Thrasher featured in *Ripley's Believe It or Not!* was credited with having a repertoire of 2,400 distinctly different songs. A popular mnemonic for their song is *Drop it drop it, cover it cover it, pull it up pull it up*. Birds nesting in shrubs or on the ground often lose young to predation; young Brown Thrashers leave, fully feathered, just 9 days after hatching, minimizing their vulnerability. Thrashers take frequent dust baths. They often pick up ants from the ground and smear or place them in their breast feathers.

two whitish wing bars

long rufous tail

heavily streaked below

||||||||||||||||||||||||||||||||||||||||||||||||||||||||||||||||||||||||||||

Curve-billed Thrasher

Toxostoma curvirostre L 11" (28 cm)

A double whistle given by both sexes in many contexts is this bird's most common call.

KEY FACTS

Most common thrasher in the Southwest. Robust bird with orange eyes and heavy, decurved bill. Dingy gray-brown plumage is spotted or mottled below. Long tail is tipped with white.

+ voice: Melodic series of low trills and warbles. Call is a sharp *whit-WHEET.*

+ habitat: Arid brush and cactus-rich desert.

+ food: Insects, also seeds, cactus fruit.

This bird is well adapted to many arid habitats.

When William Swainson examined and named a specimen of this thrasher sent from Mexico in 1827, other thrashers with curved bills hadn't yet been described for science. Two populations of this species, one in the Sonoran Desert of Arizona and one in southern Texas and New Mexico brushland, may represent different species. The song doesn't include as many imitations or repetitions as songs of mockingbirds or Brown Thrashers. Pairs defend their territory year-round. They construct their loosely woven cup nest from thorny twigs. The chicks remain in the nest for 14 to 18 days after hatching.

thick, curved bill

round spots below

long tail with white tips

European Starling

Sturnus vulgaris L 8½" (22 cm)

Hundreds of starlings can fly in tight units, maneuvering with breathtaking precision.

KEY FACTS

Stocky, short-tailed bird with pointed bill. Plumage is glossy black in summer, speckled with white in fall and winter. Juvenile is grayish brown.

+ **voice:** Elaborate song full of rattles, buzzes, and squeals.

+ **habitat:** Very common around human structures; avoids dense forest and unbroken desert.

+ **food:** Insects, berries, seeds.

Birds exhale less moisture than mammals do.

Starlings didn't live in America until 1890, when a group trying to introduce to America every bird mentioned by Shakespeare released 60 in New York's Central Park. In 1891, they released another 40. From that small beginning, more than 200 million starlings now cover the continent. Their success came directly at the expense of several declining birds, because starlings compete aggressively for nest cavities. They also are a nuisance in cities and agricultural areas. These relatives of mynas are excellent mimics, and their songs also include several sounds innate to their species. Starlings are strong fliers, moving along at about 35 to 50 mph.

triangular wings

overall grayish brown

yellow bill

glossy plumage

juvenile

breeding ♂

Cedar Waxwing

Bombycilla cedrorum L 7¼" (18 cm)

Waxwings are extremely sociable year-round, often nesting in synchrony in clustered groups.

KEY FACTS

Sleek plumage of warm browns and yellows. Brown crest and black "bandit's mask." White undertail coverts and waxy red tips to some wing feathers.

+ **voice:** Soft, high-pitched whistle *zeee*.

+ **habitat:** Widespread in areas with fruiting trees; flocks are some-what nomadic.

+ **food:** Small fruits and berries; some insects.

A waxwing plucks a berry, lets go, and gulps it.

Waxwings have a charming habit of passing petals and berries down a long line of birds until one finally eats it. Wildlife rehabbers have observed that some tough-skinned berries pass through their digestive tract intact unless the rehabber manipulates the berries to soften them; passing food among a flock may make berries more digestible and may reinforce group bonds. Many pair off in April, but nesting is delayed until fruit is plentiful. The red tips on wing feathers that give waxwings their name are more numerous on older birds. Wax-wings may use them as a signal of the reproductive value of potential mates.

juvenile

streaked underparts

crest

red, waxy tips

yellow tail tip

Snow Bunting

Plectrophenax nivalis L 6¾" (17 cm)

Snow Buntings and related longspurs were
once classified with sparrows, but aren't related.

KEY FACTS

Brown, black, and
white plumage during
winter. By spring,
brown fringes have
worn off to reveal a
starkly black-and-white
bird that soon migrates
far to the north.

+ voice: Whistled *tew*
and a musical rattle.

+ habitat: Winters
in fields, dunes, lake-
shores, often in large
flocks. Breeds in the
Arctic.

+ food: Seeds,
insects.

Snow Buntings are nicknamed "snowflakes."

In late winter, these hardy birds snow-bathe on hard-crusted snow
to abrade their dark feather tips, revealing their striking breeding
colors. Males begin returning to the high Arctic in early April, competing for territories when harsh weather is still likely. Females return
weeks later. Snow Buntings nest in skulls, cracks in large rocks, bar-
rels, construction rubble, and so on, and the best
sites are very limited. After the young are
independent, they start forming flocks.
Adults and young molt and build up
fat before migrating. Birds
in the back of a flying flock
constantly overtake
the ones in front, pro-
ducing a rolling effect.

white head

black
back

black-and-
white wings

breeding ♂

winter ♂

warm buff tones
in winter

||||||||||||||||||||||||||||||||||||||||||||||||||||||||||||||||||||||||||||

Black-and-white Warbler

Mniotilta varia L 5¼" (13 cm)

This warbler probes bark for insects, sharing many habits with creepers and nuthatches.

KEY FACTS

Entire bird is striped with alternating lines of black and white. Breeding male has black throat and cheeks; these areas are white in female and immature in fall.

+ voice: Series of high, thin *wee-see* notes.

+ habitat: Eastern warbler; common summer resident in mixed woodlands.

+ food: Insects, spiders.

This bird creeps up, down, and sideways on trees.

This aptly named bird is one of the first warblers to arrive on its breeding grounds in spring. Like many warblers, it feeds and sings in trees but nests on or near the ground, at the base of a tree or fallen log. Its high-pitched song sounds a bit like a squeaky gate and is fairly easy for bird-watchers to recognize. During migration, it associates in feeding flocks with chickadees and other songbirds by day, making its long migration flights at night. Some fly nonstop over the Gulf of Mexico to reach the Yucatán Peninsula, but some winter in the Southeast and in Texas.

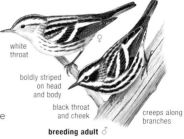

white throat

♀

boldly striped on head and body

black throat and cheek

creeps along branches

breeding adult ♂

Common Yellowthroat

Geothlypis trichas L 5" (13 cm)

This species varies in appearance over its range, and includes several subspecies.

KEY FACTS

Adult male has a broad black mask bordered above by gray or white, below by bright yellow throat and breast. Female lacks black mask and is duller yellow below.

+ **voice:** Loud, rolling song *wichity wichity wichity wich.*

+ **habitat:** Common summer resident in marshes, shrubs, grassy fields.

+ **food:** Insects, spiders.

Males have a unique face mask and golden throat.

A loud *wichity wichity wichity wich,* punctuated with harsh call notes, is a characteristic sound of marshes and wet meadows. The songster, much smaller than his voice, can stay frustratingly hidden, but patience is rewarded when he hops to a conspicuous perch. Male yellowthroats sometimes feed their mate, and both parents equally share the feeding of nestlings. When the chicks fledge, the female may start a second nest, leaving care of the first brood to her mate. Yellowthroats seldom associate with other warblers. Like other nocturnal migrants, they call while flying in the dark, probably helping maintain a safe distance from others.

broad black mask

yellow throat and breast

adult ♂

lacks mask

olive above; paler below

♀

‖‖

American Redstart

Setophaga ruticilla L 5¼" (13 cm)

The popular redstart is known as *la candelita* (the little flame) in the West Indies.

KEY FACTS

Often fans tail and spreads wings. Male is glossy black and orange; female has similar pattern in olive and yellow.

+ voice: High, thin notes, often accented at end *tsee tsee tsee tsway*; call is rich *chip*.

+ habitat: Common summer resident in second-growth woodlands.

+ food: Insects, spiders, some small fruits.

Female redstarts flash yellow tail and wing patches.

The redstart has bristly feathers on the sides of the mouth that may help funnel in insects caught on the wing. Males don't assume their black-and-orange plumage until late in their second summer, after breeding season. These yearling males claim territories and sing, but females only consider them as mates when no adult males are available. After chicks fledge, parents may divide broods and go their separate ways, each feeding their half for up to a month longer. Ornithologists placed this distinctive species in its own genus until recent studies found it to be closely related to many other warblers.

often spreads tail

glossy black and orange

gray head

adult ♂

♀

yellow patches

Yellow Warbler

Setophaga petechia L 5" (13 cm)

Bird-watchers always delight in seeing this warbler with brilliant, butter-yellow plumage.

KEY FACTS

Plump warbler with a short tail and a prominent dark eye. Adult males and females are bright yellow below; male has prominent, reddish breast streaks. Immatures are duller.

+ voice: Song is rapid *sweet sweet sweet I'm so sweet.*

+ habitat: Widespread summer resident. Common in wet habitats.

+ food: Insects.

Birds often see their reflection as a competitor.

This species is one of the most common cowbird hosts. Cowbird eggs are too large for their small bills to pierce or eject. When one is detected, the warbler may desert, or may build a new nest floor above the eggs; scientists discovered one six-tiered nest containing 11 cowbird eggs. This approach essentially throws the baby out with the bathwater, because the warbler's own eggs are lost along with the cowbird egg. If a cowbird egg doesn't appear until the warbler has two or more eggs in the nest, she's more likely to care for it. Yellow Warblers have lived more than 11 years.

adult ♂

red streaks on breast

prominent dark eye

♀

mostly yellow

Yellow-rumped Warbler

Setophaga coronata L 5½" (14 cm)

Large flocks of these warblers may be joined
by other species during winter and migration.

KEY FACTS

All have yellow rumps
and yellow side
patches. "Myrtle War-
blers" from the North
and East have white
throats; "Audubon's
Warblers" from the
West have yellow
throats. Birds are duller
and browner in fall.

+ **voice:** Song is a
simple warble.

+ **habitat:** Breeds in
coniferous and mixed
woods.

+ **food:** Insects; many
berries in winter.

Spring males are exceptionally handsome.

This abundant warbler is a generalist, wintering from the central states to Panama, and exploiting more food sources in
more habitats than other warblers. It gleans insects from leaves
and the ground, and sometimes hover-feeds. Its digestive
system is specially adapted to hold hard-to-digest
items like seeds in the gut longer than
more easily digested foods. It can also
digest the wax in bayberries, allowing it to arrive earlier in
spring, depart later in fall,
and spend the
winter farther
north than
other warblers.

widespread
subspecies

**"Myrtle
Warbler"**

breeding ♂

white
throat

**"Audubon's
Warbler"**

breeding ♂

yellow
rump

yellow
throat

brownish
upperparts

fall ♀

found in
the West

More Wood-Warblers

These "spritely butterflies of the bird world," as birding pioneer Roger Tory Peterson described them, are one of the greatest delights of birding. Nonbirders hardly know these tiny, active birds exist, as you need binoculars to see them well. Myriad patterns and a symphony of colors paint the plumage of the 50-plus species that live in North America. On these two pages, the males of an additional 24 species are shown.

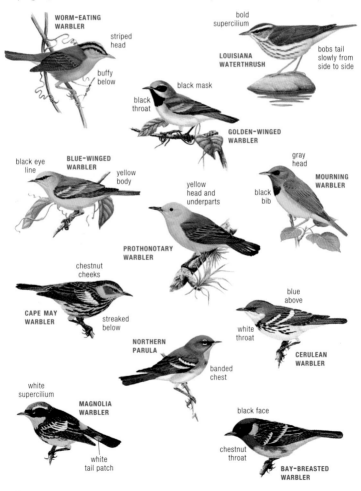

WORM-EATING WARBLER
striped head
buffy below

bold supercilium

LOUISIANA WATERTHRUSH
bobs tail slowly from side to side

black mask
black throat

GOLDEN-WINGED WARBLER

black eye line

BLUE-WINGED WARBLER
yellow body

yellow head and underparts

gray head

MOURNING WARBLER
black bib

PROTHONOTARY WARBLER

chestnut cheeks

CAPE MAY WARBLER
streaked below

blue above

white throat

NORTHERN PARULA

CERULEAN WARBLER

banded chest

white supercilium

MAGNOLIA WARBLER

white tail patch

black face

chestnut throat

BAY-BREASTED WARBLER

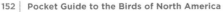

BLACKBURNIAN WARBLER
orange throat
white patch

black cap and white cheek
BLACKPOLL WARBLER
streaked below

yellow crown
CHESTNUT-SIDED WARBLER
chestnut sides

chestnut cap
PALM WARBLER
yellow below

dark blue above
BLACK-THROATED BLUE WARBLER
white wing patch

PINE WARBLER
dark cheek and yellow throat
white belly

YELLOW-THROATED WARBLER
black triangle on face
yellow throat

white stripes on face
BLACK-THROATED GRAY WARBLER
yellow spot
gray back

dark cheek
TOWNSEND'S WARBLER
yellow breast with streaks

yellow face
HERMIT WARBLER
white below

greenish cheek
BLACK-THROATED GREEN WARBLER

gray back
red and black head
RED-FACED WARBLER

YELLOW-BREASTED CHAT
bright yellow below
large size

Eastern Towhee

Pipilo erythrophthalmus L 7½" (19 cm)

Towhees share their ground-foraging techniques with their junco and sparrow relatives.

KEY FACTS

Male has black hood and upperparts, rufous sides; female is chocolate brown where the male is black. Long tail. In the West, very similar Spotted Towhee has white spots above.

+ voice: Song is ringing *drink-your-tea!* Call is up-slurred *towhee.*

+ habitat: Found in tangles, thickets, overgrown fields.

+ food: Insects, seeds, berries.

Most Eastern Towhees have brilliant red eyes.

In 1585, John White painted a male and female towhee during a visit to the short-lived settlement on Roanoke Island, bringing the bird to the attention of European ornithologists. Towhees eat, sing, and roost at eye level or below, often noisily scratching at the ground, moving leaves aside with a two-footed backward hop. They are difficult to see in the dense foliage where they spend most of their time. Light-eyed towhees live in Florida and often skulk behind Florida Scrub-Jays, especially when people are feeding the jays, perhaps to steal food from the jays' caches. This once abundant species has steadily declined in the Northeast.

milk chocolate upperparts

♀

black upperparts

white tail spots obvious in flight

rufous sides

♂

California Towhee

Melozone crissalis L 9" (23 cm)

This towhee allows close study; it's so sedentary that marked birds are easy to track.

KEY FACTS

Plain brown plumage with some orange-buff color around face and under tail. Scratches in leaf litter or on bare ground for food.

+ voice: Series of simple *chink* notes.

+ habitat: Resident in chaparral and scrubland from Baja California to Oregon.

+ food: Mostly seeds, some insects.

Rusty undertail coverts provide a splash of color.

plain brownish overall

long tail

blurry streaks

juvenile

buffy on throat and face

This common and well-known bird is tolerant of urbanization and not particularly shy around human observers. Mated pairs remain together for life, both chasing any other towhees off their defended territory, and the two are usually found in very close proximity. During the breeding season, mated pairs produce squealing duets about three times every hour—these duets coordinate their breeding behaviors. Unexpectedly, over 40 percent of all nests contain "extra-pair" young—those with a different father or mother than the mated pair. This and the similar-looking Canyon Towhee found east of the Colorado River were once considered the same species, the "Brown Towhee."

Chipping Sparrow

Spizella passerina L 5½" (14 cm)

The "Chippy," named for its calls, is a summer breeder in much of North America.

KEY FACTS

Slender sparrow with long tail and dark line through eye in all plumages. Breeding adult has chestnut cap and gray face; winter adult has streaked chestnut crown; first-winter bird has buff breast.

+ **voice:** Long trill of dry *chip* notes.

+ **habitat:** Mix of trees and grassy openings. Northern breeders migrate.

+ **food:** Insects, seeds.

The rusty cap and black eye lines are distinctive.

rufous crown
blackish eye line extends to bill
gray nape
breeding adult
gray rump
1st winter
long tail

This handsome backyard sparrow has adapted to many human-modified habitats. In spring, males arrive in many areas just as juncos are moving on. Females arrive a week or two later. The female builds the compact nest, usually in a conifer and often at eye level, as soon as weather patterns are favorable. She lines the nest with horsehairs and fur plucked from resting horses, sleeping dogs, and other animals. Males sometimes feed their incubating mate, and share in feeding the young. An Ontario study revealed that after a male has attracted a mate, he may wander through neighboring territories, mating with other females.

|||||||||||||||||||||||||||||||||||||||||||||||||||||||||||||||||||||||||||||||||||||||

Field Sparrow

Spizella pusilla L 5¾" (15 cm)

In spring, the male sings with gusto, but after
finding a mate, he sings much less often.

KEY FACTS

Long-tailed sparrow
that is closely related
to Chipping Sparrow
(previous page), but
has pale head with
prominent white eye
ring and pink bill.

+ **voice:** Series of
clear, plaintive whistles
that slowly accelerate
into a trill.

+ **habitat:** Fairly
common in overgrown
fields and open, brushy
woodlands; found east
of the Rockies.

+ **food:** Insects, seeds.

Field Sparrows have long tails and plain faces.

Field Sparrows live in brushy pastures and second growth, but
shy away from similar habitat in suburbs and other settled areas.
They seldom associate in flocks. The female builds the first nest
of the season using grasses, on or near the ground at the base of
woody vegetation. Later nests, as ground cover grows taller, are
built in small saplings and shrubs. White-tailed deer eat the eggs
and nestlings of ground-nesting sparrows far more than was sus-
pected. Field Sparrows were probably
most abundant in the 1800s, after the
eastern forest was cleared but before
widespread development. They
have declined significantly
since the Breeding Bird
Survey began in 1966.

western birds
are grayer
overall

pink
bill

distinct
white
eye ring

buffy breast

Song Sparrow

Melospiza melodia L 4¾–6¾" (13–17 cm)

Thoreau described the song as *Maids! Maids! Maids! Hang up your teakettle-ettle-ettle.*

KEY FACTS

Song Sparrows vary in overall coloration and size—small and pale in the desert to large and dark in Alaska—but all are russet and gray birds with bold streaks on the chest that often form a central spot.

+ voice: Series of notes with a trill in the middle or as a final flourish.

+ habitat: Widespread and common, especially in brushy areas.

+ food: Insects, seeds.

Birds lose little body heat via their legs and feet.

This handsome, streaked sparrow nests in many suburban and even urban neighborhoods but is often overlooked, its pretty but quiet songs blending into background noise. Males communicate with neighboring males by song, and sing most persistently throughout the day while trying to attract a mate. One May day in her Ohio backyard, Margaret Morse Nice followed an 8-year-old marked male who had just lost his mate, counting his songs. Beginning at 4:44 a.m. (36 minutes before sunrise), he sang 2,305 songs that day. At least one Song Sparrow banded as an adult in Colorado was still alive when re-trapped in Colorado over 11 years later.

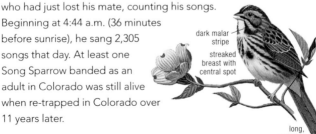

dark malar stripe

streaked breast with central spot

long, rounded tail

||||||||||||||||||||||||||||||||||||||||||||||||||||||||||||||||||||||||||||||

White-crowned Sparrow

Zonotrichia leucophrys L 7" (18 cm)

The sparrows in the genus *Zonotrichia* are big, with bold markings and distinctive songs.

KEY FACTS

Adult has thick black-and-white crown stripes; immature has brown and light tan stripes. Back is striped with rusty brown and pale gray, and underparts are mostly pale gray.

+ voice: Mournful whistles followed by a trill.

+ habitat: Breeds in mountains or far to the north; favors brushy edges in winter.

+ food: Seeds, insects.

Young White-crowns are often misidentified.

This handsome sparrow has a pleasing song, heard occasionally in winter and more often during spring migration. Throughout the year and throughout its range, this sparrow is found near grass, bare ground, and dense shrubs or conifers; it feeds mostly on the ground. Easy to maintain in captivity and observe in the wild, the White-crown is one of the most thoroughly studied of all birds. Thanks to it, scientists have learned more about evolutionary biology, how birds learn their songs, how songs vary geographically, migration, physiology, and many other subjects.

orange (to yellow or pinkish) bill

reddish brown and tan crown stripes

streaked back

immature

black-and-white crown stripes

adult

White-throated Sparrow

Zonotrichia albicollis L 6¾" (17 cm)

Like avian chipmunks, these stripe-headed
sparrows eat seeds on the ground.

KEY FACTS

Medium-size sparrow
with white throat and
bright yellow spot
near eye. Two color
morphs—white-striped
and tan-striped—
describe the color of the
stripe above the eye.

+ voice: Mournful,
whistled song is heard
year-round.

+ habitat: Breeds in
the North; very com-
mon in East in winter;
frequents bird feeders,
brush, woodland edges.

+ food: Seeds, insects.

The yellow "lore" spot is a good field mark.

This sparrow's clear, whistled
*Old Sam Peabody, Peabody,
Peabody* is heard in backyards
as well as in northern forests
where it breeds. Half of both
males and females have white head
stripes, the other half tan stripes; each

tan supercilium

**tan-striped
morph**

dark
bill

white throat
and supercilium

yellow
spot in front
of eye

**white-striped
morph**

prefers the opposite pattern in a mate, as if blond humans always
selected a dark-haired mate and vice versa. About 96 percent of all
pairs include one of each color. White-striped birds of both sexes
sing and are territorially aggressive. Tan-striped birds of both sexes
provide excellent parental care. Though tan-striped males are less
aggressive than white-striped, their territories in an area tend to be
the same size.

Dark-eyed Junco

Junco hyemalis L 6¼" (16 cm)

This distinctive sparrow produces a shorter, more musical trill than the Chipping Sparrow.

KEY FACTS

Juncos show a bewildering amount of variation. Eastern "Slate-colored" is mostly gray with a white belly; western "Oregon" has a black-ish hood and rusty back. All have white outer tail feathers.

+ voice: Song is a simple trill.

+ habitat: Breeds in mountains or in the north; winter flocks favor fields and edges.

+ food: Seeds, insects.

"Oregon Juncos" are a western group of juncos.

Our familiar "snowbird" has been a popular backyard bird since colonial days. When juncos fly away, white outer tail feathers alert others of danger. A high-protein diet before molt increases the amount of white, so the color also serves as a signal to potential mates that the bird may be a good provider. Juncos are territorial during the breeding season. In winter, they gather in huge flocks that mill about on the ground like "little gray-robed monks and nuns," as ornithologist Florence Merriam Bailey wrote in 1899.

all types have white outer tail feathers

"Slate-colored" ♂
dark gray hood and white belly

widespread type
many plumage variations depending on location

"Gray-headed"
southern Rockies type
rufous back

More Towhees and Sparrows

Over 50 species of sparrows and related birds can be found in North America. They have the reputation of being "difficult," difficult to see well and difficult to identify. Granted, many of them are streaky LBJs (little brown jobs) that hide in bushes, but a close inspection will sort them out. Head pattern is often the best clue, but bill color and the presence or absence of streaked underparts are also useful.

spotted back

black head

SPOTTED TOWHEE

rufous flanks

♂

rufous crown

CANYON TOWHEE

buffy throat

breast spot

ABERT'S TOWHEE

warm brown overall

blackish face

cinnamon undertail coverts

rufous head and back streaks

buffy below

BACHMAN'S SPARROW

rufous patch at breast sides

AMERICAN TREE SPARROW

winter

breast spot

white eye ring

bold head pattern

gray nape

VESPER SPARROW

chestnut wing patch

long tail

fall

CLAY-COLORED SPARROW

white outer tail feathers

distinctive head pattern

breast spot

LARK SPARROW

extensive white in tail

gray head and white eye ring

SAGE SPARROW

black stripe

breast spot

mostly black

white wing patch

LARK BUNTING

breeding ♂

yellowish supercilium

SAVANNAH SPARROW

heavily streaked

short tail

reddish streaks on back

buffy breast

GRASSHOPPER SPARROW

short tail

broad gray supercilium

fine streaks on buffy breast

LINCOLN'S SPARROW

reddish crown

rufous wings

grayish breast

breeding

SWAMP SPARROW

black crown and bib

HARRIS'S SPARROW

pink bill

breeding

large size

yellow crown and black eyebrow

GOLDEN-CROWNED SPARROW

gray face

breeding

Scarlet Tanager

Piranga olivacea L 7" (18 cm)

The tanagers of North America, in the cardinal family, aren't related to tropical tanagers.

KEY FACTS

Breeding male has red body and glossy black wings; female is olive above and yellowish below. In fall, male resembles female.

+ voice: Robin-like, raspy song. Call is a harsh *chip-burr*.

+ habitat: Eastern hardwood forests; winters in South America.

+ food: Insects, fruits.

Breeding males are stunningly brilliant.

This exceptionally vivid bird hides in plain sight in eastern deciduous forests due to its shy, secretive ways. In South America in winter and during migration, tanagers associate in flocks, but when they arrive on their breeding grounds, each male immediately establishes and defends a territory. The song consists of long, raspy but musical phrases often likened to a robin with a sore throat. Females return about a week after males; nesting begins soon after. After the young fledge, the parents stay with them for a couple of weeks before the family disperses. Males undergo a dramatic molt in late summer, but are seldom seen during this time.

♀

olive green back

yellowish underparts

glossy black wings and tail

brilliant red

breeding adult ♂

Western Tanager

Piranga ludoviciana L 7¼" (18 cm)

This tanager was first collected for science by
Lewis and Clark in Idaho.

KEY FACTS

Breeding adult male
has a red head;
mostly yellow body;
and black wings, tail,
and back. Female is
olive-gray above and
yellowish below. Both
sexes have conspicu-
ous wing bars.

+ **voice:** Robin-like,
raspy song. Call is a
pit-er-ick.

+ **habitat:** Western
pine forests; winters
in Mexico, Central
America.

+ **food:** Insects, fruits.

This western bird breeds as far north as Alaska.

One of the most strikingly
beautiful of all North
American birds, this inhabitant
of coniferous forests ranges
farther north than any
other tanager. Its red
color is produced from
a pigment found in very few birds. Rhodoxanthin is
found in conifer needles, and tanagers probably get
it from eating insects that feed on the needles. Mate selection
may take place during winter or on migration. Even during the
breeding season, Western Tanagers aren't very aggressive, but
do escort intruders off their territory. Rare sightings in the East
are usually at feeders.

reddish
head

**breeding
adult ♂**

gray
"saddle"

whitish
wing
bars

black
"saddle" and
yellow rump

♀

yellowish
underparts

Northern Cardinal

Cardinalis cardinalis L 8¾" (22 cm)

Many sports teams are named for this bird, the most popular state bird of all.

KEY FACTS

Male is instantly recognizable by red color, long crest, black face, and triangular red bill. Female is tawny brown with red accents; similar juvenile has black bill.

+ voice: Call is a sharp *chip*. Songs include a loud *cheer cheer cheer*.

+ habitat: Resident in the East.

+ food: Insects, seeds; common at bird feeders.

Winter cardinals can be very sociable at feeders.

Which is more beautiful: a cardinal's plumage or voice? Bright color indicates the quality of a cardinal's diet. Redder males and females with brighter underwings provide more food to their chicks; those males defend higher quality territories, too. Unexpectedly, captive females don't show a preference for brighter mates. Both sexes sing. Females incubate the eggs, and males provide more food for the nestlings. Cardinals are nonmigratory, but many disperse in fall. The range has crept northward since the early 1800s. The seven states that named the cardinal state bird form an interconnected block.

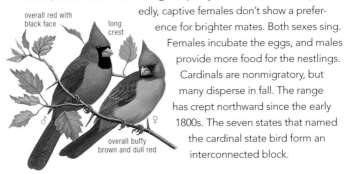

overall red with black face

long crest

overall buffy brown and dull red

||||||||||||||||||||||||||||||||||||||||||||||||||||||||||||||||||||||||||||||||||

Rose-breasted Grosbeak

Pheucticus ludovicianus L 8" (20 cm)

This bird is nicknamed "cutthroat" for its plumage, and "potato bug" bird for its diet.

KEY FACTS

Adult male is black and white, with a rose-colored breast and huge pale bill. Female looks like a big finch—brown above, streaked below, with two wing bars.

+ voice: Rich warbling song. Call is a very sharp *eek!*

+ habitat: Eastern forests and edges; winters in Central America, Cuba.

+ food: Insects, fruit, seeds, buds.

Both parents feed fledglings.

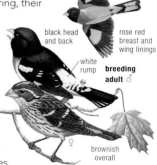

black head and back

rose red breast and wing linings

white rump

breeding adult ♂

♀

brownish overall

From the time grosbeaks arrive in spring, their striking appearance and song make them very welcome at feeders. Females have a subtle beauty; spring males lack any trace of subtlety. Gaudy as they are, males incubate the eggs for hours each day, often singing while on the nest. Both parents feed the young. Like cardinal chicks, fledglings hop about but cannot fly for a few days, usually staying hidden in foliage. Families remain together until migration. They hybridize with the Black-headed Grosbeak where their ranges overlap in the Great Plains. Hybrid females produce smaller clutches, and more of their eggs don't hatch.

Indigo Bunting

Passerina cyanea L 5½" (14 cm)

Based on DNA studies, buntings belong to the same colorful family as the cardinal.

KEY FACTS

Small, compact song-bird. Breeding male is electric blue but can look black from a distance; female is rich brown with faint breast streaks.

+ **voice:** Cheerful song of paired phrases *sweet-sweet chew-chew sweet-sweet.* Call is sharp *spit.*

+ **habitat:** Summer resident in thickets, weedy fields.

+ **food:** Insects, seeds, buds.

Breeding pairs don't spend much time together.

During the breeding season, male Indigo Buntings sing from conspicuous perches, continuing later in the afternoon and longer into summer than most songbirds. They don't learn their father's song; rather, when they're on their own first territory, they learn their song by interacting with neighboring males. Females, which raise the chicks with little help, are quiet and secretive, and seem to have no preference between males who sing the same song as nearby birds and those singing a different tune. They may raise three or four broods in summer before retreating to the tropics for winter. A few winter in Florida.

deep blue plumage

brown upperparts

silvery bill

breeding adult ♀

breeding adult ♂

blurry streaks below

||||||||||||||||||||||||||||||||||||||||||||||||||||||||||||||||||||||||||||||||||||

Painted Bunting

Passerina ciris L 5½" (14 cm)

When skulking in vegetation, even the gaudiest males can be surprisingly hard to see.

KEY FACTS

Adult male is an amazing mix of vivid colors—deep blue, red, and lime green. Female is bright green above, yellow-green below; juvenile is grayer.

+ voice: Song is rapid series of short phrases. Call is a loud *chip*.

+ habitat: Summer resident of southern thickets and edges.

+ food: Insects, spiders, seeds.

Males may sing from one perch for many minutes.

adult ♂

unmistakable

bright green upperparts

yellow-green below

no wing bars ♀

This "most wanted" species on many birders' lists is called the *nonpareil* (without equal) in French. There are two separate populations. Western birds make a short postbreeding migration to Arizona and parts of Mexico to molt, and then continue on to Mexico and beyond. Eastern birds molt on their breeding grounds before migrating to the Caribbean. These nocturnal migrants sometimes alight in exhaustion on ships in the Gulf of Mexico. Young males resemble females until their second autumn, when their brilliant adult feathers grow in. Their beauty entices people in the winter range to capture many for the pet trade, a factor in the species' decline since the 1960s.

Western Meadowlark

Sturnella neglecta L 9½" (24 cm)

Many members of the blackbird family, such as meadowlarks and orioles, have bright plumage.

KEY FACTS

Chunky, robin-size songbird with flat head and long, slender bill. Bright yellow breast is crossed with black "V," and upperparts are cryptic mix of brown and tan. In flight, note the white outer tail feathers.

+ voice: Series of bubbling, flutelike notes; sharp *chuck* note.

+ habitat: Grasslands and agricultural fields.

+ food: Probes soil for insects, grain, seeds.

Fluffed feathers insulate birds during cold weather.

In 1805, Lewis and Clark collected "a kind of larke" in Montana and described differences in tail and song from the Eastern Meadowlark. Audubon, taken with its song and mystified about why it had been overlooked since the expedition, gave it the scientific name *neglecta*, but it wasn't accepted as a separate species until 1910. The Eastern Meadowlark's simple whistle lacks the universal appeal of the Western's rich, bubbling song—the Western is state bird of six states, the Eastern of none. Meadowlarks sing on perches around their territory's border, often along roadsides. They expose prey by inserting their bill into soil and forcing it open.

yellow submoustachial (unlike Eastern Meadowlark)

black "V" on chest

yellow throat and underparts

stocky and short tailed

Red-winged Blackbird

Agelaius phoeniceus L 8¾" (22 cm)

When America's most abundant, familiar blackbird first appears in marshes, spring is here.

KEY FACTS

Stocky, short-tailed bird that forages on the ground. Adult male is glossy black with vivid red shoulders; female is dark brown above and heavily streaked below.

+ **voice:** Song is a gurgling *konk-la-ree*; call is a flat *chack*.

+ **habitat:** Marshes, meadows; huge winter flocks congregate in agricultural areas.

+ **food:** Seeds, grain, insects.

Females look like oversize sparrows.

Males arrive on breeding marshes before females and immediately start displaying, singing exuberantly while exposing their red epaulets to perfection. Those with the best displays and territories may attract as many as 15 females. DNA tests of nestlings indicate that females often mate with more than one male as well. Among the most abundant species on the continent, Red-wings form tremendous flocks during migration and winter, often damaging crops. Millions are killed each year via shooting, trapping, pesticides, and spraying roosts with wetting agents during cold weather. Humans provide both their primary sources of food and causes of mortality.

pale supercilium

adult ♀

strongly streaked below

adult ♂

red shoulders most visible when singing

Yellow-headed Blackbird

Xanthocephalus xanthocephalus L 9½" (24 cm)

This and many other blackbirds nest colonially and flock together the rest of the year.

KEY FACTS

Brilliant yellow head and black body of adult male are unmistakable. Female has buffy yellow breast; head and upperparts are dusky brown.

+ voice: Song ends in a long, strangled buzz; call is a rich *croak*.

+ habitat: Prairie wetlands and western marshes; flocks winter in agricultural areas.

+ food: Insects, seeds, grain.

The male's displays are fun to see and hear.

This bird's breeding range is centered on prairie wetlands, often in the same marshes as Red-wings. Yellow-heads dominate, claiming territories over the deepest water with cattails and bulrushes. They nest in association with some terns, cooperatively attacking predators. People liken the Yellow-head's song to the grating noise of a rusty hinge and may call it ugly, bizarre, weird, or hilarious; females find it irresistible. Experiments reveal that if a Red-wing's epaulets are blackened, he loses his territory, but if a Yellow-head's head is blackened, he can retain his territory, usurp another's territory, and attract mates.

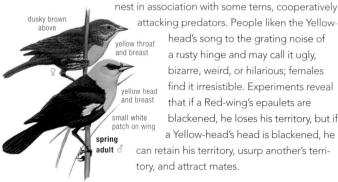

dusky brown above

yellow throat and breast

♀

yellow head and breast

small white patch on wing

spring adult ♂

Brewer's Blackbird

Euphagus cyanocephalus L 9" (23 cm)

This "miniature grackle" has such lustrous plumage that it's been called the "satin bird."

KEY FACTS

Male is black with purple gloss on head, green gloss on body. Female is flat, brownish gray with dark eyes; male's eyes are yellow.

+ **voice:** Wheezy, unmusical *que-ee*; call is a harsh *check*.

+ **habitat:** Open areas of the West, from grasslands to city sidewalks.

+ **food:** Insects, seeds, grain.

Males have yellow eyes; females' are dark.

Formerly only found in the West, Brewer's Blackbird was not recorded nesting east of the Great Plains until 1914, when it spread east in a dramatic range expansion. Newly arriving blackbirds supplanted Common Grackles in wild and rural areas, but the grackles won the competition in cities and towns. In turn, as grackles spread west, they started replacing Brewer's Blackbirds in cities and towns but not grasslands. Some birds breeding in Canadian prairies migrate west, wintering in the coastal regions of British Columbia and Washington. Where abundant, they are an important food source for falcons, yet at least one banded bird in California survived over 12 years.

pale yellow eye

glossy black plumage

♂

dark eye

overall dull gray-brown

♀

Common Grackle

Quiscalus quiscula L 12½" (32 cm)

Grackles, like many birds, smear ants into their feathers, possibly to control parasites.

KEY FACTS

Blackbird with flared tail and pale eyes. Iridescent plumage varies from purplish in South and most of East Coast to bronzy with blue head. Female has nonglossy, brown body.

✛ **voice:** Short, creaky *readle-eak;* call is a deep *chuck.*

✛ **habitat:** Open woods, farmland, marshes, suburbs.

✛ **food:** Insects, seeds, grain.

Blackbirds fly straight compared to many birds.

Right when the first robins arrive, grackles suddenly appear, males puffed up in territorial displays, strutting about, bills pointed skyward. When they fly, they lower and extend the central tail feathers to form an impressive keel. They forage on lawns, devouring slugs, cutworms, and other pests. During nesting, they become secretive. About half the males remain with their mates during incubation, and often help feed the noisy young. Grackles raid nests and attack small birds—perhaps what led Ogden Nash to "deem the grackle / an ornithological debacle."

blackish with purple gloss to head

"Purple Grackle" ♂

keel-shaped tail

"Bronzed Grackle" ♂

yellow eye

bluish head contrasts with bronze-green body

||||||||||||||||||||||||||||||||||||||||||||||||||||||||||||||||||||||||||||||||||||||

Great-tailed Grackle

Quiscalus mexicanus ♂L 18" (46 cm) ♀L 15" (38 cm)

Until 1983, this and the coastal Boat-tailed
Grackle were considered the same species.

KEY FACTS

Very large grackle
with pale eyes. Purple-
glossed male has
extremely long keel-
shaped tail; brownish
female is smaller and
shorter-tailed, with
buffy throat.

+ voice: Loud series
of whistles, trills, and
ratchet-like sounds.

+ habitat: Farms,
feedlots, wetlands,
lawns, and urban areas.

+ food: Insects,
small animals, grain,
garbage.

This crow-size blackbird has a streamlined look.

It's impossible to
ignore breeding
colonies of these flam-
boyant birds with their whistles,
squawks, screeches, bugle-like
calls, and soft tinkling notes, which
can continue all night at roosts near

large grackle
with small head
and long bill
♂

glossy
black
overall

buffy throat
and breast
♀

shorter
tail

very long,
keel-shaped
tail

artificial lighting. Displaying males may engage in fierce
battles. Only the largest ones with the longest tails hold territories,
each containing one to several trees. Several females may construct
nests and raise young in the same tree. At all ages more females than
males survive. A rapid range expansion northward during the last
century, coinciding with irrigation and urbanization, has brought this
formerly tropical and subtropical species as far north as Minnesota.

Brown-headed Cowbird

Molothrus ater L 7½" (19 cm)

Two other parasitic cowbirds are expanding their ranges in the United States.

KEY FACTS

Small, compact blackbird with short tail and dark eyes. Male is glossy black with rich brown head. Female is plain, gray-brown; juvenile is also brown, but streaked below and scaly above.

+ voice: Male song is a liquid gurgling; calls include rattles and chattering.

+ habitat: Woodlands, farmlands, suburbs.

+ food: Seeds, grain, some insects.

Cowbirds often associate with cattle . . . or bison.

"Buffalo birds" were once limited to short-grass plains following bison, feeding on insects where the heavy-hoofed giants cut into prairie sod. Widespread extirpation of bison coincided with plowing the prairies and cattle ranching; rather than disappearing, cowbirds increased and spread. Females search for nests of small songbirds in which to lay their eggs—they produce as many as 40 eggs each year. Females check on these nests, often trashing one if their egg has been removed. Some parasitized species raise a cowbird or two and their own young successfully; others have declined significantly since cowbirds arrived.

brown head contrasts with black body

♂

short tail

gray-brown plumage

often feeds with tail cocked up

♀

||||||||||||||||||||||||||||||||||||||||||||||||||||||||||||||||||||||||||||||||||||||||||||

Hooded Oriole

Icterus cucullatus L 8" (20 cm)

Orioles belong to the genus *Icterus*, which means jaundiced, for the birds' yellow color.

A favorite nest site is under a palm frond.

KEY FACTS

Slender, long-tailed oriole. Male is deep yellow or orange with a black bib, back, and tail. Female is olive above, yellow below; first-spring male is similar, but has black bib.

+ voice: Calls include loud, whistled *wheet* and rapid chattering.

+ habitat: Summer resident, usually nests in shade trees and palms; winters in Mexico.

+ food: Insects, spiders, fruits, nectar.

Quieter than other orioles, this beautiful bird expanded its range as ornamental plants and feeders became popular. Those breeding in northern Mexico and the United States were once entirely migratory, but now many remain in southern California and Texas year-round. In California, they're sometimes called the "palm leaf oriole" for building their suspended nests in fan palm leaves. Rio Grande Valley birds once depended on Spanish moss for nest construction, but today, nests are more often woven of palm leaf fibers. They are a primary host for Bronzed Cowbirds.

black bib

yellow-orange and black plumage

breeding adult ♂

♀ yellowish underparts

black throat

yellow head and underparts

1st spring ♂

Baltimore Oriole

Icterus galbula L 8¼" (21 cm)

The state bird of Maryland was named for the colors on Lord Baltimore's family crest.

KEY FACTS

Stocky, short-tailed oriole. Brilliant orange and black male can be hard to see in the treetops; female is paler orange below, brownish olive above; head can be blackish.

+ voice: Series of rich, fluting whistles; staccato chatter call.

+ habitat: Summer resident of eastern woodlands.

+ food: Insects, fruit, nectar; jelly feeders, cut fruit.

Migrating orioles often visit orange feeders.

This striking bird visits feeding stations for oranges and sugar water. It nests in tall shade trees, especially near water. The female anchors the pouch-like nest at the fork of twigs too slender to support most predators. Sometimes, adults and nestlings become entangled in long nest fibers, so people are cautioned never to provide string or yarn longer than 6 to 8 inches. Baltimore Orioles molt into fresh plumage before migrating to the tropics in fall. Scattered individuals may winter in northern states. From 1973 to 1994, this and the western Bullock's Oriole were considered to be the same species.

maximum black

spring adult ♀

black hood

1st spring ♀

breeding adult ♂

bright orange rump and underparts

American Goldfinch

Spinus tristis L 5" (13 cm)

With coast-to-coast popularity, this is the state bird of Washington, Iowa, and New Jersey.

KEY FACTS

Small finch with short tail. Breeding male is bright yellow with black cap; breeding female is olive-brown above, yellow below. Winter male is grayish brown below with yellow face, tan upperparts.

+ **voice:** Twitters, trills, *su-wee* notes; flight call *per-chik-o-ree.*

+ **habitat:** Widespread in overgrown fields, pastures, suburbs.

+ **food:** Mostly seeds; common feeder visitor.

Goldfinches specialize in feeding on seeds.

The "wild canary" is one of the few vegans among birds, even feeding nestlings regurgitated seeds. It undergoes a complete molt of body feathers both in early spring and in fall. It waits longer than other birds to begin nesting, both because it takes time to recover from the spring molt and because goldfinches use thistle seeds for nest construction and food. Cowbird nestlings don't survive long when raised by goldfinches because of the low-protein diet. Some goldfinches winter quite far north; others move long distances. One banded in March in Ontario was recovered, dead, in Louisiana, 1,000 miles away, 8 months later.

yellow-brown upperparts

winter adult ♂

white undertail coverts

olive above; pale yellow below

breeding ♀

black cap

bright yellow body

breeding ♂

female less rufous above

House Finch

Carpodacus mexicanus L 6" (15 cm)

In backyards and apartment balconies, this popular bird often nests in hanging baskets.

KEY FACTS

Sparrow-size finch with a stubby bill, curved on top. Male has red eyebrow, breast, and rump (sometimes yellow). Female and juvenile are grayish brown with streaked underparts.

+ **voice:** Lively warble call is whistled *wheat.*

+ **habitat:** Widespread in city parks, farms, backyards, and forest edges.

+ **food:** Seeds, buds, fruit; visits feeders.

Both parents feed their young regurgitated food.

Originally a western species, House Finches were once trapped and sold as "Hollywood Finches." In 1940, to avoid prosecution under the Migratory Bird Act, unscrupulous caged-bird dealers released captive finches on Long Island. The birds established a population that spread along the coast and westward. In 1994, an epidemic of a common poultry eye disease struck in the mid-Atlantic states, decimating local populations and rapidly spreading north and west. The disease was especially deadly in eastern populations, probably in part because eastern birds lacked genetic diversity, having descended from a handful of birds.

broad reddish eyebrow

red throat and breast

♂

indistinct face pattern

lightly streaked overall ♀

House Sparrow

Passer domesticus L 6¼" (16 cm)

Unrelated to sparrows native to America, this is the sparrow mentioned in the Bible.

KEY FACTS

Big-headed sparrow with heavy bill and short tail. Male has black bib, gray crown, and chestnut nape; pattern less evident in winter. Female has buffy eyebrow and streaked back.

+ voice: Series of *chirrup* notes.

+ habitat: Abundant in human-altered habitats, especially cities, towns, and farms.

+ food: Seeds, grain, insects; visits feeders.

These sociable birds are usually seen in groups.

buffy eyebrow

plain underparts; streaked back

♀

gray crown

chestnut nape

black bib

breeding ♂

subdued head pattern

fall ♂

This artful dodger has attached itself to civilization throughout history. Immigrants who missed the familiar European bird released captives many times in the 1800s, most famously in New York City around 1850. People claimed the sparrows would devour insect pests, but they mostly eat grains. Their rapid spread sparked declines in native birds, agricultural losses, and long-standing debates among ornithologists. Currently, they are slowly decreasing in America and more rapidly in their native Europe. In winter, they roost in buildings, outside lights, and near other heat sources. Some fly underneath parked cars to warm themselves until the engine cools.

About the Authors

LAURA ERICKSON has been avidly birding since 1975. She has authored six books about birds—including *Sharing the Wonder of Birds with Kids,* winner of the National Outdoor Book Award—writes for birding magazines, and blogs for the American Birding Association. She has been a licensed wildlife rehabilitator and previously served as science editor at the Cornell Lab of Ornithology. Her long-running radio program, *For the Birds,* airs on public and community radio stations, and is podcast via iTunes. She lives in Duluth, Minnesota, with her husband and an education screech-owl, Archimedes.

JONATHAN ALDERFER, artist, author, and editor, has contributed paintings to the *National Geographic Field Guide to the Birds of North America,* and co-authored the sixth edition in 2011. He has written or edited numerous bird books for National Geographic, most recently *Backyard Guide to the Birds of North America* and *Bird-watcher's Bible.* A selection of his original art can be viewed at jonathanalderfer.com. He and his wife live in Washington, D.C.

Further Resources

BOOKS

Alderfer, Jonathan, ed. *Complete Birds of North America.* National Geographic Society, 2005.

Alderfer, Jonathan, and Jon L. Dunn. *Birding Essentials.* National Geographic, 2007.

Alderfer, Jonathan, and Paul Hess. *Backyard Guide to the Birds of North America.* National Geographic, 2011.

Dunn, Jon L., and Jonathan Alderfer, eds. *Field Guide to the Birds of North America,* 6th edition. National Geographic Society, 2011.

Dunne, Pete. *Pete Dunne's Essential Field Guide Companion.* Houghton Mifflin, 2006.

Elbroch, Mark, and Eleanor Marks. *Bird Tracks & Sign.* Stackpole Books, 2001.

Erickson, Laura. *The Bird Watching Answer Book.* Workman Publishing Company, 2009.

Erickson, Laura. *101 Ways to Help Birds.* Stackpole Books, 2006.

Kaufman, Kenn. *Lives of North American Birds.* Houghton Mifflin, 1996.

Kroodsma, Donald. *The Backyard Birdsong Guide.* Chronicle Books, 2008.

Roth, Sally. *Bird-by-Bird Gardening.* Rodale, 2006.

Sibley, David A. *The Sibley Guide to Birds.* Alfred A. Knopf, 2000.

Stokes, Donald and Lillian. *The Stokes Field Guide to the Birds of North America.* Little, Brown and Company, 2010.

MAGAZINES

Bird Watcher's Digest: birdwatchersdigest.com
Birding: aba.org/birding/
BirdWatching: birdwatchingdaily.com
Living Bird: allaboutbirds.org/page.aspx?pid=1173
North American Birds: aba.org/nab/
Western Birds: westernfieldornithologists.org

WEBSITES

American Birding Association: *aba.org*
Birding News: *birding.aba.org*
Christmas Bird Count: *birds.audubon.org/christmas-bird-count*
Cornell Lab of Ornithology: *birds.cornell.edu*
eBird: *ebird.org*
Jonathan Alderfer: *jonathanalderfer.com*
Laura Erickson's For the Birds: *lauraerickson.com*
National Geographic Society: *nationalgeographic.com/animals/birding*
Surfbirds: *surfbirds.com*
Virtual Birder: *virtualbirder.com*

Acknowledgments

I am indebted to the American Ornithologists' Union and the Cornell Lab of Ornithology for many of the resources that I consulted while working on this book. In particular, the online publication of *The Birds of North America* has been an invaluable source of detailed and trustworthy information on North America's breeding birds. Two ornithologists I met early in my birding career, Chandler Robbins and Joe Grzybowski, have ever been generous with information and warm in their encouragement. My co-author, Jonathan Alderfer, inspired me with his discerning eye, broad knowledge, and long track record of excellent work. My husband, Russell, and our three children have always been exceptionally supportive and understanding, whether I was rehabbing a duck in the bathtub or dragging them along on a detour to see the birds at a sewage pond.

—LAURA ERICKSON

Most of the Key Facts information in this book is based on the comprehensive species accounts in National Geographic's *Field Guide to the Birds of North America,* authored primarily by Jon Dunn. The bird illustrations were originally painted for that same field guide—my congratulations to the artists on the accuracy and artistry of those images. Bob Steele expertly guided us to some of the best bird photographers in North America. Credits for both the illustrations and the photographs can be found on the next page. It was a great pleasure to work with my co-author, Laura Erickson, whose good-natured approach to birding and depth of knowledge shines through on every page. My wife, Zora Margolis Alderfer, supplied editorial advice and made numerous suggestions that greatly improved the readability of the Key Facts section that I authored— thank you for your guidance and unflagging support.

—JONATHAN ALDERFER

Illustrations Credits

PHOTOS: Cover: (UP), Maxis Gamez/www.gvisions.org; LO (Left to Right), Prem Balson/NG My Shot; Scott Evers/NG My Shot; John Reasbeck/NG My Shot; Laura Mountainspring/NG My Shot; Spine, Al Mueller/Shutterstock; Back Cover: (Left to Right), Steve Byland/Shutterstock; George Chiu/NG My Shot; Bret Douglas/NG My Shot; Howard Cheek/NG My Shot; 2-3, Jim Zipp; 7, RobertMcCaw.com; 9, erlre74/Shutterstock; 11, Alan Murphy; 12, Michel Lamarche/NG My Shot; 13, Oleg Slyusarchuk/NG My Shot; 14, Robert Royse; 15, Tim Zurowski; 16, Bert De Tilly/NG My Shot; 17, Homer Caliwag; 18, Marie Read; 19, E.J. Peiker; 20, David Hemmings; 21, E.J. Peiker; 24, Michael Ohaion/NG My Shot; 25, Bob Gress; 26, Rob Kemp/Shutterstock; 27, Phillip Holland/Shutterstock; 28, Richard Walsh/NG My Shot; 29, Sylvie Goguen/NG My Shot; 30, James Cumming/NG My Shot; 31, Marie Read; 32, Homer Caliwag; 33, Jim Neiger/www.flightschoolphotography.com; 34, Danny Brown/NG My Shot; 35, E.J. Peiker; 36, Kevin T. Karlson; 37, E.J. Peiker; 38, Steve Ellwood/NG My Shot; 39, Mia McPherson/onthewingphotography.com; 40, Jim Zipp; 41, Larsek/Shutterstock; 42, Harry Moulis/NG My Shot; 43, Robert Royse; 44, Bob Steele; 45, Bret Douglas/NG My Shot; 46, Clay Billman; 47, Kenneth Rush/Shutterstock; 48, Alan Murphy; 49, Kevin T. Karlson; 50, Jim Neiger/www.flightschoolphotography.com; 51, Lloyd Spitalnik; 52, David Hemmings; 53, Jim Zipp; 56, RobertMcCaw.com; 57, David March/NG My Shot; 58, Maxis Gamez/www.gvisions.org; 59, Lloyd Spitalnik; 60, Jim Zipp; 61, RobertMcCaw.com; 62, Richard Fitzer/Shutterstock; 63, Homer Caliwag; 64, Kevin T. Karlson; 65, Bob Steele; 66, E.J. Peiker; 67, Alan Murphy; 68, Bob Steele; 69, Lloyd Spitalnik; 70, RobertMcCaw.com; 71, Alan Murphy; 72, Homer Caliwag; 73, Glenn Bartley; 74, Homer Caliwag; 75, Bob Steele; 76, Lloyd Spitalnik; 77, E.J. Peiker; 80, Pooja C. Raghavendra/NG My Shot; 81, Bob Steele; 82, Jim Zipp; 83, Alan Murphy; 84, Lloyd Spitalnik; 85, Alan Murphy; 86, Alan Murphy; 87, Deborah Smith/NG My Shot; 88, Alan Murphy; 89, Robert Postma/NG My Shot; 90, David Hemmings; 91, Paul Ayick/NG My Shot; 92, Bob Steele; 93, Judd Patterson; 94, Terence P. Brashear; 95, Alan Murphy; 96, Brian E. Small; 97, Bob Steele; 98, Glenn Bartley; 99, Alan Murphy; 100, FloridaStock/Shutterstock; 102, Judd Patterson; 103, Steve Byland/Shutterstock; 104, Brian E. Small; 105, Jack Sutton/NG My Shot; 106, Brook Burling/NG My Shot; 107, Eileen Worman/NG My Shot; 108, John Stankewitz/NG My Shot; 109, Judd Patterson; 110, Mia McPherson/onthewingphotography.com; 111, Terence P. Brashear; 112, Steven Smith/NG My Shot; 113, Tony Campbell/Shutterstock; 114, Bob Steele; 115, RobertMcCaw.com; 116, Julie Bishop/NG My Shot; 117, Patricia Fiedler/NG My Shot; 118, Alan Murphy; 119, Mark McMaster/NG My Shot; 120, Steve Hamilton/NG My Shot; 121, Bob Gress; 122, David Hemmings; 123, Marie Read; 124, Steven Smith/NG My Shot; 125, Roger van Gelder; 126, Peter Friedlieb/NG My Shot; 127, Jaromir Penicka/NG My Shot; 128, Roger van Gelder; 129, Alexander Viduetsky/NG My Shot; 130, RobertMcCaw.com; 131, Bill Dalton/NG My Shot; 132, Jim Zipp; 133, Alan Murphy; 134, RobertMcCaw.com; 135, Jeff Brubaker/NG My Shot; 136, Jeff Pinkerton/NG My Shot; 137, Gary Hamilton/NG My Shot; 138, Rowland Willis/NG My Shot; 139, Beth Pirie/NG My Shot; 140, Martha Marks/Shutterstock; 141, Alan Murphy; 142, Lloyd Spitalnik; 143, Eirini Pajak/NG My Shot; 144, Erik Mandre/NG My Shot; 145, George Chiu/NG My Shot; 146, Marie Read; 147, Lloyd Spitalnik; 148, Michael G. Mill/Shutterstock; 149, Alan Murphy; 150, RobertMcCaw.com; 151, Laura L. Erickson;

154, RobertMcCaw.com; 155, Brian E. Small; 156, Marie Read; 157, Robert Royse; 158, Richard Cronberg/NG My Shot; 159, David Seibel; 160, Garth McElroy; 161, Alan Murphy; 164, Lloyd Spitalnik; 165, Bob Steele; 166, Roy Orr/NG My Shot; 167, Brian Lasenby/Shutterstock; 168, Danny Brown/NG My Shot; 169, Alan Murphy; 170, Stacey Huston/NG My Shot; 171, Steve Creek/NG My Shot; 172, Judd Patterson; 173, Dennis Donohue/Shutterstock; 174, Roger van Gelder; 175, Beverly Cochran/NG My Shot; 176, Jose Hernandez/NG My Shot; 177, RobertMcCaw.com; 178, Mark Lewer/NG My Shot; 179, Matthew Studebaker; 180, Stubblefield Photography/Shutterstock; 181, turtix/Shutterstock; 182 (UP), Laura L. Erickson; 182 (LO), Mark Thiessen, NGP.

BIRD ARTWORK: Back Cover (Western Tanager and Bald Eagle), Jonathan Alderfer; 1, Jonathan Alderfer; 5, Cynthia J. House; 6, John C. Pitcher; 10, Jonathan Alderfer; 12-23, Cynthia J. House; 24-28, Kent Pendleton; 29, David Quinn; 30, Jonathan Alderfer; 31, H. Jon Janosik; 32, Jonathan Alderfer; 33 & 34, H. Jon Janosik; 35 (UPLE & UPRT), H. Jon Janosik; 35 (LO), Jonathan Alderfer; 36-38, Diane Pierce; 39 (UP), Thomas R. Schultz; 39 (LO), Diane Pierce; 40, Peter Burke; 41-43, Diane Pierce; 44-46, Donald L. Malick; 47, N. John Schmitt; 48, Donald L. Malick; 49 (UPRT), N. John Schmitt; 49 (LOLE), Donald L. Malick; 50 & 51, N. John Schmitt; 52 & 53, Donald L. Malick; 54-5, All artwork on these pages by Donald L. Malick, Kent Pendleton, and N. John Schmitt; 56 & 57, Marc R. Hanson; 58, Diane Pierce; 59, Jonathan Alderfer; 60, John C. Pitcher; 61, Killian Mullarney; 62, H. Jon Janosik; 63 & 64, John C. Pitcher; 65, Michael O'Brien; 66, Daniel S. Smith; 67, John C. Pitcher; 68 (UPLE), Daniel S. Smith; 68 (LOLE & LORT), Thomas R. Schultz; 69, John C. Pitcher; 70, Jonathan Alderfer; 71-76, Thomas R. Schultz; 77, Chuck Ripper; 78-9, All artwork on these pages by Jonathan Alderfer, H. Jon Janosik, Killian Mullarney, Michael O'Brien, and Thomas R. Schultz; 80, H. Douglas Pratt; 81-84, N. John Schmitt; 85 & 86, H. Douglas Pratt; 87-91, Donald L. Malick; 92, Thomas R. Schultz; 93 & 94, N. John Schmitt; 95-98, H. Douglas Pratt; 99-103, Donald L. Malick; 104, David Beadle; 105-107, H. Douglas Pratt; 108, Jonathan Alderfer; 109-113, H. Douglas Pratt; 114, N. John Schmitt; 115, H. Douglas Pratt; 116, N. John Schmitt; 117 (LE), N. John Schmitt; 117 (RT), Jonathan Alderfer; 118, David Beadle; 119 (UP & CTR), H. Douglas Pratt; 119 (LO), Thomas R. Schultz; 120-122, H. Douglas Pratt; 123 & 124, Michael O'Brien; 125, John P. O'Neill; 126 & 127, H. Douglas Pratt; 128, Thomas R. Schultz; 129 (UP), Thomas R. Schultz; 129 (LO), H. Douglas Pratt; 130 & 131, H. Douglas Pratt; 132 & 133, N. John Schmitt; 134-137, H. Douglas Pratt; 138, Thomas R. Schultz; 139 & 140, H. Douglas Pratt; 141 (UP), N. John Schmitt; 141 (LO), H. Douglas Pratt; 142, H. Douglas Pratt; 143, N. John Schmitt; 144 & 145, H. Douglas Pratt; 146, Diane Pierce; 147 & 148, Thomas R. Schultz; 149, H. Douglas Pratt; 150, Thomas R. Schultz; 151 (UPRT & LOLE), H. Douglas Pratt; 151 (LORT), Thomas R. Schultz; 152-153, All artwork on these pages by David Beadle, Peter Burke, H. Douglas Pratt, and Thomas R. Schultz; 154 & 155, Peter Burke; 156, Thomas R. Schultz; 157 & 158, Diane Pierce; 159, Thomas R. Schultz; 160, Diane Pierce; 161 (UP), N. John Schmitt; 161 (CTR & LO), Diane Pierce; 162-163, All artwork on these pages by David Beadle, Peter Burke, and Diane Pierce; 164 & 165, Peter Burke; 166 & 167, Diane Pierce; 168-170, Thomas R. Schultz; 171-175, H. Douglas Pratt; 176, N. John Schmitt; 177 & 178, Peter Burke; 179, Diane Pierce; 180, Thomas R. Schultz; 181, N. John Schmitt.

Index

Boldface indicates species profile.

National Geographic Pocket Guide to the Birds of North America

Laura Erickson and Jonathan Alderfer

CELEBRATING

‹125›

YEARS

Published by the National Geographic Society

John M. Fahey, *Chairman of the Board and Chief Executive Officer*

Timothy T. Kelly, *President*

Declan Moore, *Executive Vice President; President, Publishing and Digital Media*

Melina Gerosa Bellows, *Executive Vice President; Chief Creative Officer, Books, Kids, and Family*

Prepared by the Book Division

Hector Sierra, *Senior Vice President and General Manager*

Janet Goldstein, *Senior Vice President and Editorial Director*

Jonathan Halling, *Design Director, Books and Children's Publishing*

Marianne R. Koszorus, *Design Director, Books*

Susan Tyler Hitchcock, *Senior Editor*

R. Gary Colbert, *Production Director*

Jennifer A. Thornton, *Director of Managing Editorial*

Susan S. Blair, *Director of Photography*

Meredith C. Wilcox, *Director, Administration and Rights Clearance*

Staff for This Book

Jonathan Alderfer, *Project Editor*

Sanaa Akkach, *Art Director*

Linda Makarov, *Designer*

Bob Steele, *Photography Editor*

Carl Mehler, *Director of Maps*

Paul Lehman, *Chief Map Researcher and Editor*

Sven M. Dolling, *Map Production Manager*

Marshall Kiker, Michael O'Connor, *Associate Managing Editors*

Judith Klein, *Production Editor*

Mike Horenstein, *Production Manager*

Galen Young, *Illustrations Specialist*

Katie Olsen, *Production Design Assistant*

Erin Stone, *Production Intern*

Margaret Murphy, *Design Intern*

Manufacturing and Quality Management

Phillip L. Schlosser, *Senior Vice President*

Chris Brown, *Vice President, NG Book Manufacturing*

George Bounelis, *Vice President, Production Services*

Nicole Elliott, *Manager*

Rachel Faulise, *Manager*

Robert L. Barr, *Manager*

The National Geographic Society is one of the world's largest nonprofit scientific and educational organizations. Founded in 1888 to "increase and diffuse geographic knowledge," the Society's mission is to inspire people to care about the planet. It reaches more than 400 million people worldwide each month through its official journal, *National Geographic,* and other magazines; National Geographic Channel; television documentaries; music; radio; films; books; DVDs; maps; exhibitions; live events; school publishing programs; interactive media; and merchandise. National Geographic has funded more than 10,000 scientific research, conservation and exploration projects and supports an education program promoting geographic literacy. For more information, visit www.nationalgeographic.com.

For more information, please call 1-800-NGS LINE (647-5463) or write to the following address:

National Geographic Society
1145 17th Street N.W.
Washington, D.C. 20036-4688 U.S.A.

For information about special discounts for bulk purchases, please contact National Geographic Books Special Sales: ngspecsales@ngs.org

For rights or permissions inquiries, please contact National Geographic Books Subsidiary Rights: ngbookrights@ngs.org

Library of Congress Cataloging-in-Publication Data

Erickson, Laura, 1951-
  National Geographic pocket guide to the birds of North America / Laura Erickson, Jonathan Alderfer.
    pages cm
  Includes bibliographical references and index.
  ISBN 978-1-4262-1044-0 (alk. paper)
  1. Birds--North America--Identification. I. Alderfer, Jonathan K. II. National Geographic Society (U.S.) III. Title. IV. Title: Pocket guide to the birds of North America. V. Title: Birds of North America.
  QL681.E74 2012
  598.097--dc23

                                        2012034597

Printed in Hong Kong

13/THK/2

FIVE FIELD GUIDES IN ONE!

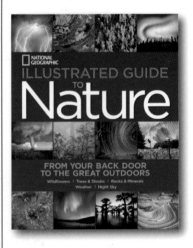

Bursting with beautiful images and illustrations, as well as tips to help you identify different species in the wild, this comprehensive nature field guide is a great tool for families, amateur adventurers, and empty nesters who want to enjoy and understand nature—from wilderness camping to your own backyard!

- ■ **800 natural wonders from the U.S and Canada**
- ■ **Gorgeous photographs**
- ■ **Text from experts in the field**
- ■ **Easy-to-read yet authoritative**

Chapters include:
Wildflowers
Trees & Shrubs
Rocks & Minerals
Weather
Night Sky

Also Available ·······················

 Like us on Facebook.com: Nat Geo Books

 Follow us onTwitter.com: @NatGeoBooks

NATIONAL GEOGRAPHIC

AVAILABLE WHEREVER BOOKS ARE SOLD
nationalgeographic.com/books